THE FACTS ON FILE
CALCULUS
HANDBOOK

ELI MAOR, Ph.D.
Adjunct Professor of Mathematics,
Loyola University, Chicago, Illinois

Facts On File, Inc.

I dedicate this book to the countless students who,
over the past 300 years,
had to struggle with the intricacies of the differential
and integral calculus—and prevailed.
You have my heartiest congratulations!

The Facts On File Calculus Handbook
Copyright © 2003 by Eli Maor, Ph.D.

Facts On File
132 West 31st Street
New York NY 10001

Library of Congress Cataloging-in-Publication Data

Maor, Eli.
    The Facts On File calculus handbook / Eli Maor.
        p. cm.
Includes bibliographical references and index.
    ISBN 0-8160-4581-X (acid-free paper)
    1. Calculus—Handbooks, manuals, etc. I. Title.
QA303.2.M36 2003
515—dc21   2003049027

Facts On File books are available at special discounts when purchased in bulk quantities for businesses, associations, institutions, or sales promotions. Please call our Special Sales Department in New York at 212/967-8800 or 800/322-8755.

You can find Facts On File on the World Wide Web at
http://www.factsonfile.com

Cover design by Cathy Rincon
Illustrations by Anja Tchepets and Kerstin Porges

Printed in the United States of America

MP Hermitage 10 9 8 7 6 5 4 3 2 1

This book is printed on acid-free paper.

CONTENTS

PREFACE

Over the past 25 years or so, the typical college calculus textbook has grown from a modest 350-page book to a huge volume of some 1,200 pages, with thousands of exercises, special topics, interviews with career mathematicians, 10 or more appendixes, and much, much more. But as the old adage goes, more is not always better. The enormous size and sheer volume of these monsters (not to mention their weight!) have made their use a daunting task. Both student and instructor are lost in a sea of information, not knowing which material is important and which can be skipped. As if the study of calculus is not a challenge already, these huge texts make the task even more difficult.

The Facts On File Calculus Handbook is an attempt to come to the student's rescue. Intended for the upper middle school, high school, and college students who are taking a single-variable calculus class, this will be a quick, ideal reference to the many definitions, theorems, and formulas for which the subject is notorious.

The reader will find important terms listed alphabetically in the Glossary section, accompanied by illustrations wherever relevant. Most entries are supplemented by at least one example to illustrate the concept under discussion.

The Biographies section has brief sketches of the lives and contributions of many of the men and women who played a role in bringing the calculus to its present state. Other names, such as Euclid or Napier, are also included because of their overall contribution to mathematics and science in general. The Chronology section surveys the development of calculus from its early roots in ancient Greece to our own times.

Section four lists the most-frequently used trigonometric identities, a selection of differentiation and integration formulas, and a summary of the various convergence tests for infinite series. Finally, a Recommended Reading section lists many additional works in calculus and related areas of interest, thus allowing the reader to further expand his or her interest in the subject.

In compiling this handbook, I gave practicality and ease of use a high priority, putting them before scholarly pedantry. For example, when discussing a function, I have used both the notations f and $y = f(x)$, although, from a purely pedantic point of view there is a difference between the two (the former is the name of the function, while the latter denotes the number that f assigns to x).

It is my hope that *The Facts On File Calculus Handbook,* together with Facts On File's companion handbooks in algebra and geometry, will provide mathematics students with a useful aid in their studies and a valuable supplement to the traditional textbook. I wish to thank Frank K. Darmstadt, my editor at Facts On File, for his valuable guidance in making this handbook a reality.

THE CALCULUS: A HISTORICAL INTRODUCTION

The word *calculus* is short for *differential and integral calculus;* it is also known as the *infinitesimal calculus.* Its first part, the differential calculus, deals with change and rate of change of a function. Geometrically, this amounts to investigating the *local* properties of the graph that represents the function—those properties that vary from one point to another. For example, the *rate of change* of a function, or in geometric terms, the slope of the tangent line to its graph, is a quantity that varies from point to point as we move along the graph. The second part of the calculus, the integral calculus, deals with the *global* features of the graph—those properties that are defined for the entire graph, such as the area under the graph or the volume of the solid obtained by revolving the graph about a fixed line. At first thought, these two aspects of the calculus may seem unrelated, but as Newton and Leibniz discovered around 1670, they are actually inverses of one another, in the same sense that multiplication and division are inverses of each other.

It is often said that Sir Isaac Newton (1642–1727) in England and Gottfried Wilhelm Leibniz (1646–1716) in Germany invented the calculus, independently, during the decade 1665–75, but this is not entirely correct. The central idea behind the calculus—to use the limit process to obtain results about graphs, surfaces, or solids—goes back to the Greeks. Its origin is attributed to Eudoxus of Cnidus (ca. 370 B.C.E.), who formulated a principle known as the *method of exhaustion.* In Eudoxus's formulation:

> *If from any magnitude there be subtracted a part not less than its half, from the remainder another part not less than its half, and so on, there will at length remain a magnitude less than any preassigned magnitude of the same kind.*

By "magnitude" Eudoxus meant a geometric construct such as a line segment of given length. By repeatedly subtracting smaller and smaller parts from the original magnitude, he said, we can make the remainder as small as we please—*arbitrarily* small. Although Eudoxus formulated his principle verbally, rather than with mathematical symbols, it holds the germ of our modern "ε-δ" definition of the limit concept.

The first who put Eudoxus's principle into practice was Archimedes of Syracuse (ca. 287–212 B.C.E.), the legendary scientist who defeated the Roman fleet besieging his city with his ingenious military inventions (he was reportedly

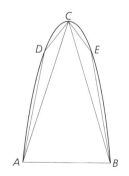

Area of a parabolic segment

slain by a Roman soldier while musing over a geometric theorem which he drew in the sand). Archimedes used the method of exhaustion to find the area of a sector of a parabola. He divided the sector into a series of ever-smaller triangles whose areas decreased in a geometric progression. By repeating this process again and again, he could make the triangles fit the parabola as closely as he pleased—"exhaust" it, so to speak. He then added up all these areas, using the formula for the sum of a geometric progression. In this way he found that the total area of the triangles approached 4/3 of the area of the triangle ABC. In modern language, the combined area of the triangles approaches the *limit* 4/3 (taking the area of triangle ABC to be 1), as the number of triangles increases to infinity. This result was a great intellectual achievement that brought Archimedes within a hair's breadth of our modern integral calculus.

Why, then, didn't Archimedes—or any of his Greek contemporaries—actually discover the calculus? The reason is that the Greeks did not have a working knowledge of algebra.

To deal with infinite processes, one must deal with variable quantities and thus with algebra, but this was foreign to the Greeks. Their mathematical universe was confined to geometry and some number theory. They thought of numbers, and operations with numbers, in geometric terms: a number was interpreted as the length of a line segment, the sum of two numbers was the combined length of two line segments laid end-to-end along a straight line, and their product was the area of a rectangle with these line segments as sides. In such a static world there was no need for variable quantities, and thus no need for algebra. The invention of calculus had to wait until algebra was developed to the form we know it today, roughly around 1600.

In the half century preceding Newton and Leibniz, there was a renewed interest in the ancient method of exhaustion. But unlike the Greeks, who took great care to wrap their mathematical arguments in long, verbal pedantry, the new generation of scientists was more interested in practical results. They used a loosely defined concept called "indivisibles"—an infinitely small quantity which, when added infinitely many times, was expected to give the desired result. For example, to find the area of a planar shape, they thought of it as made of infinitely many "strips," each infinitely narrow; by adding up the areas of these strips, one could find the area in question, at least in principle. This method, despite its shaky foundation, allowed mathematicians to tackle many hitherto unsolved problems. For example, the astronomer Johannes Kepler (1571–1630), famous for discovering the laws of planetary motion, used indivisibles to find the volume of various solids of revolution (reportedly he was led to this by his dissatisfaction with the way wine merchants gauged the volume of wine in their casks). He thought of each solid as a collection of infinitely many thin slices, which he then summed up to get the total volume.

Many mathematicians at the time used similar techniques; sometimes these methods worked and sometimes they did not, but they were always cumbersome and required a different approach for each problem. What was needed was a unifying principle that could be applied to *any* type of problem with ease and efficiency. This task fell to Newton and Leibniz.

Newton, who was a physicist as much as a mathematician, thought of a function as a quantity that continuously changed with time—a "fluent," as he called it; a curve was generated by a point P(x, y) moving along it, the coordinates x and y continuously varying with time. He then calculated the *rates of change* of x and y with respect to time by finding the difference, or change, in x and y between two "adjacent" instances, and dividing it by the elapsed time interval. The final step was to let the elapsed time become infinitely small or, more precisely, to make it so small as to be negligible compared to x and y themselves. In this way he expressed each rate of change as a function of time. He called it the "fluxion" of the corresponding fluent with respect to time; today we call it the *derivative.*

Once he found the rates of change of x and y with respect to time, he could find the rate of change of *y with respect to x.* This quantity has an important geometric meaning: it measures the steepness of the curve at the point P(x, y) or, in other words, the slope of the tangent line to the curve at *P.* Thus Newton's "method of fluxions" is equivalent to our modern *differentiation*—the process of finding the derivative of a function y = f(x) with respect to *x.* Newton then formulated a set of rules for finding the derivatives of various functions; these are the familiar *rules of differentiation* which form the backbone of the modern calculus course. For example, the derivative of the sum of two functions is the sum of their derivatives [in modern notation $(f + g)' = f' + g'$], the derivative of a constant is zero, and the derivative of a product of two functions is found according to the *product rule* $(fg)' = f'g + fg'$. Once these rules were formulated, he applied them to numerous curves and successfully found their slopes, their highest and lowest points (their maxima and minima), and a host of other properties that could not have been found otherwise.

But that was only half of Newton's achievement. He next considered the *inverse* problem: given the fluxion, find the fluent, or in modern language: given a function, find its *antiderivative.* He gave the rules for finding antiderivatives of various functions and combinations of functions; these are today's *integration rules.* Newton then turned to the problem of finding the area under a given curve; he found that this problem and the tangent problem (finding the slope of a curve) are *inverses* of each other: in order to find the area under a graph of a function f, one must first find an antiderivative of f. This inverse relation is known as the *Fundamental Theorem of Calculus,* and it unifies the two branches of the calculus, the differential calculus and the integral calculus.

Across the English Channel, Leibniz was working on the same ideas. Although Newton and Leibniz maintained cordial relations, they were working independently and from quite different points of view. While Newton's ideas were rooted in physics, Leibniz, who was a philosopher at heart, followed a more abstract approach. He imagined an "infinitesimal triangle" formed by a small portion of the graph of f, an increment Δx in x, and a corresponding increment Δy in y. The ratio $\Delta y/\Delta x$ is an approximation to the slope of the tangent line to the graph at the point P(x, y). Leibniz thought of Δx and Δy as infinitely small quantities; today we say that the slope of the tangent line is the *limit* of $\Delta y/\Delta x$ as Δx approaches zero ($\Delta x \to 0$), and we denote this limit by dy/dx. Similarly, Leibniz thought of the area under the graph of f as the sum of infinitely many narrow strips of width Δx and heights $y = f(x)$; today we formulate this idea in terms of the limit concept. Finally, Leibniz discovered the inverse relation between the tangent and area problems.

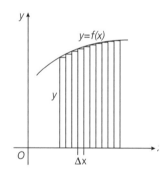

Approximating a tangent line

Thus, except for their different approach and notation, Newton and Leibniz arrived at the same conclusions. A bitter priority dispute between the two, long simmering behind the facade of cordial relations, suddenly erupted in the open, and the erstwhile colleagues became bitter enemies. Worse still, the dispute over who should get the credit for inventing the calculus would poison the academic atmosphere in Europe for more than a hundred years. Today Newton and Leibniz are given equal credit for inventing the calculus—the greatest development in mathematics since Euclid wrote his *Elements* around 300 B.C.E.

Area under a function

Knowledge of the calculus quickly spread throughout the world, and it was immediately applied to a host of problems, old and new. Among the first to be tackled were two famous unsolved problems: to find the shape of a chain of uniform thickness hanging freely under the force of gravity, and to find the curve along which a particle under the force of gravity will slide down in the shortest possible time. The first problem was solved simultaneously by Leibniz, Jakob Bernoulli of Switzerland, and the Dutch scientist Christiaan Huygens in 1691, each using a different method; the shape turned out to be the graph of $y = \cosh x$ (the hyperbolic cosine of x), a curve that became known as the *catenary* (from the Latin *catena,* a chain). The second problem, known as the *brachistochrone* (from the Greek words meaning "shortest time"), was solved in 1691 by Newton, Leibniz, the two Bernoulli brothers, Johann and

Jakob, and the Frenchman Guillaume François Antoine L'Hospital (who in 1696 published the first calculus textbook); the required curve turned out to be a *cycloid,* the curve traced by a point on the rim of a wheel as it rolls along a straight line. The solutions to these problems were among the first fruits of the newly invented calculus.

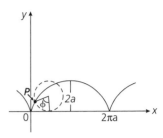

Cycloid

The 18th century saw an enormous expansion of the calculus to new areas of investigation. Leonhard Euler (1707–83), one of the most prolific mathematicians of all time, is regarded as the founder of modern *analysis*— broadly speaking, the study of infinite processes and limits. Euler discovered numerous infinite series and infinite products, among them the series $\pi^2/6 = 1/1^2 + 1/2^2 + 1/3^2 + \ldots$, regarded as one of the most beautiful formulas in mathematics. He also expanded the methods of calculus to complex variables (variables of the form $x + iy$, where *x* and *y* are real numbers and $i = \sqrt{-1}$), paving the way to the *theory of functions of complex variables,* one of the great creations of 19th-century mathematics. Another branch of analysis that received great attention during this period (and still does today) is *differential equations*—equations that contain an unknown function and its derivatives. A simple example is the equation $y' = ky$, where $y = f(x)$ is the unknown function and *k* is a constant. This equation describes a variety of phenomena such as radioactive decay, the attenuation of sound waves as they travel through the atmosphere, and the cooling of an object due to its surrounding; its solution is $y = y_0 e^{kx}$, where y_0 is the initial value of *y* (the value when $x = 0$), and *e* is the base of natural logarithms (approximately 2.7182818). The techniques for solving such equations have found numerous applications in every branch of science, from physics and astronomy to biology and social sciences.

In the 19th century the calculus was expanded to three dimensions, where solids and surfaces replace the familiar graphs in two dimensions; this *multivariable* calculus, and its extension to vectors, became an indispensable tool of physics and engineering. Another major development of the early 19th century was the discovery by Jean-Baptiste-Joseph Fourier that any "reasonably-behaved" function, when regarded as a periodic function over an interval of length *T,* can be expressed as an infinite sum of sine and cosine terms whose periods are integral divisors of *T* (see *Fourier series* in the Glossary section). These *Fourier series* are central to the study of vibrations and waves, and they played a key role in the development of quantum mechanics in the early 20th century.

But while these developments have greatly enlarged the range of problems to which the calculus could be applied, several 19th-century mathematicians felt that the calculus still needed to be put on firm, logical foundations, free from any physical or geometric intuition. Foremost among them was Augustin-Louis

Cauchy (1789–1857), who was the first to give a precise, rigorous definition of the limit concept. This emphasis on rigor continued well into the 20th century and reached its climax in the years before World War II (in 1934 Edmund Landau published a famous calculus textbook in which not a single figure appeared!). Since the war, however, the pendulum has swung back toward a more balanced approach, and the old distinction between "pure" and "applied" mathematics has largely disappeared.

Today the calculus is an indispensable tool not only in the natural sciences but also in psychology and sociology, in business and economics, and even in the humanities. To give just one example, a business owner may want to find the number of units he or she should produce and sell in order to maximize the business's profit; to do so, it is necessary to know how the cost of production C, as well as the revenue R, depend on the number x of units produced and sold, that is, the functions $C(x)$ and $R(x)$ (the former usually consists of two parts—*fixed costs,* which are independent of the number of units produced and may include insurance and property taxes, maintenance costs, and employee salaries, and *variable costs* that depend directly on x). The *Profit P* is the difference between these two functions and is itself a function of x, $P(x) = R(x) - C(x)$. We can then use the standard methods of calculus to find the value of x that will yield the highest value of P; this is the *optimal production level* the business owner should aim at.

SECTION ONE
GLOSSARY

abscissa The first number of an ordered pair (x, y); also called the *x*-coordinate.

absolute convergence *See* CONVERGENCE, ABSOLUTE.

absolute error *See* ERROR, ABSOLUTE.

absolute maximum *See* MAXIMUM, ABSOLUTE.

absolute minimum *See* MINIMUM, ABSOLUTE.

absolute value The absolute value of a real number x, denoted $|x|$, is the number "without its sign." More precisely, $|x| = x$ if $x \geq 0$, and $|x| = -x$ if $x < 0$. Thus $|5| = 5$, $|0| = 0$, and $|-5| = -(-5) = 5$. Geometrically, $|x|$ is the distance of the point x from the origin O on the number line.
    See also TRIANGLE INEQUALITY.

absolute-value function The function $y = f(x) = |x|$. Its domain is all real numbers, and its range all nonnegative numbers.

acceleration The rate of change of velocity with respect to time. If an object moves along the *x*-axis, its position is a function of time, $x = x(t)$. Then its velocity is $v = dx/dt$, and its acceleration is $a = dv/dt = d(dx/dt)/dt = d^2x/dt^2$, where d/dt denotes differentiation with respect to time.

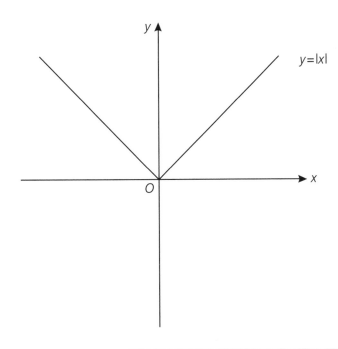

Absolute-value function

addition of functions The sum of two functions f and g, written $f + g$. That is to say, $(f + g)(x) = f(x) + g(x)$. For example, if $f(x) = 2x + 1$ and $g(x) = 3x - 2$, then $(f + g)(x) = (2x + 1) + (3x - 2) = 5x - 1$. A similar definition holds for the difference of f and g, written $f - g$.

additive properties of integrals

1. $\int_a^c f(x)\,dx + \int_c^b f(x)\,dx = \int_a^b f(x)\,dx$. In abbreviated form, $\int_a^c + \int_c^b = \int_a^b$.

Note: Usually c is a point in the interval $[a, b]$, that is, $a \le c \le b$. The rule, however, holds for any point c at which the integral exists, regardless of its relation relative to a and b.

2. $\int_a^b [f(x) + g(x)]\,dx = \int_a^b f(x)\,dx + \int_a^b g(x)\,dx$, with a similar rule for the difference $f(x) - g(x)$. The same rule also applies for indefinite integrals (antiderivatives).

algebraic functions The class of functions that can be obtained from a finite number of applications of the algebraic operations addition, subtraction, multiplication, division, and root extraction to the variable x. This includes all polynomials and rational functions (ratios of polynomials) and any finite number of root extractions of them; for example, $\sqrt{x} + \sqrt[3]{x}$.

algebraic number A zero of a polynomial function $f(x)$ with integer coefficients (that is, a solution of the equation $f(x) = 0$). All rational numbers are algebraic, because if $x = a/b$, where a and b are two integers with $b \neq 0$, then x is the solution of the linear equation $bx - a = 0$. Other examples are $\sqrt{2}$ (the positive solution of the quadratic equation $x^2 - 2 = 0$) and $\sqrt[3]{1 + \sqrt{2}}$ (a solution of the sixth-degree polynomial equation $x^6 - 2x^3 - 1 = 0$). The imaginary number $i = \sqrt{-1}$ is also algebraic, because it is the solution of the equation $x^2 + 1 = 0$ (note that in all the examples given, all coefficients are integers).

See also TRANSCENDENTAL NUMBER.

alternating *p*-series *See* p-SERIES, ALTERNATING.

alternating series *See* SERIES, ALTERNATING.

amplitude One-half the width of a sine or cosine graph. If the graph has the equation $y = a \sin(bx + c)$, then the amplitude is $|a|$, and similarly for $y = a \cos(bx + c)$.

analysis The branch of mathematics dealing with continuity and limits. Besides the differential and integral calculus, analysis includes differential equations, functions of a complex variable, operations research, and many more areas of modern mathematics.

See also DISCRETE MATHEMATICS.

analytic geometry The algebraic study of curves, based on the fact that the position of any point in the plane can be given by an ordered pair of numbers (coordinates), written (x, y). Also known as *coordinate geometry,* it was invented by Pierre de Fermat and René Descartes in the first half of the 17th century. It can be extended to three-dimensional space, where a point P is given by the three coordinates x, y, and z, written (x, y, z).

angle A measure of the amount of rotation from one line to another line in the same plane.

Between lines: If the lines are given by the equations $y = m_1x + b_1$ and $y = m_2x + b_2$, the angle between them—provided neither of the lines is vertical—is given by the formula $\phi = \tan^{-1}(m_2 - m_1)/(1 + m_1m_2)$. For example, the angle between the lines $y = 2x + 1$ and $y = 3x + 2$ is $\phi = \tan^{-1}(3 - 2)/(1 + 3 \cdot 2) = \tan^{-1} 1/7 \approx 8.13$ degrees.

Between two curves: The angle between their tangent lines at the point of intersection.

Of inclination of a line to the x-axis: The angle $\phi = \tan^{-1} m$, where m is the slope of the line. Because the tangent function is periodic, we limit the range of ϕ to $0 \leq \phi \leq \pi$.

See also SLOPE.

angular velocity Let a line through the origin rotate with respect to the x-axis through an angle θ, measured in radians in a counterclockwise sense. The angle of rotation is thought of as continuously varying with time (as the hands of a clock), though not necessarily at a constant rate. Thus θ is a function of the time, $\theta = f(t)$. The *angular velocity,* denoted by the Greek letter ω (omega), is the derivative of this function: $\omega = d\theta/dt = f'(t)$. The units of ω are radians per second (or radians per minute).

annuity A series of equal payments at regular time intervals that a person either pays to a bank to repay a loan, or receives from the bank for a previously-deposited investment.

antiderivative The antiderivative of a function f(x) is a function F(x) whose derivative is f(x); that is, $F'(x) = f(x)$. For example, an antiderivative of $5x^2$ is $5x^3/3$, because $(5x^3/3)' = 5x^2$. Another antiderivative of $5x^2$ is $5x^3/3 + 7$, and in fact $5x^3/3 + C$, where C is an arbitrary constant.

The antiderivative of f(x) is also called an *indefinite integral* and is denoted by $\int f(x)\,dx$; thus $\int 5x^2\,dx = 5x^3/3 + C$.
    See also INTEGRAL, INDEFINITE.

approximation A number that is close, but not equal, to another number whose value is being sought. For example, the numbers 1.4, 1.41, 1.414, and 1.4142 are all approximations to $\sqrt{2}$, increasing progressively in accuracy. The word also refers to the *procedure* by which we arrive at the approximated number. Usually such a procedure allows one to approximate the number being sought to any desired accuracy. Associated with any approximation is an estimate of the *error* involved in replacing the true number by its approximated value.
    See also ERROR; LINEAR APPROXIMATION

Archimedes, spiral of (linear spiral) A curve whose polar equation is $r = a\theta$, where *a* is a constant. The grooves of a vinyl disk have the shape of this spiral.

arc length The length of a segment of a curve. For example, the length of an arc of a circle with radius *r* and angular width θ (measured in radians) is $r\theta$. Except for a few simple curves, finding the arc length involves calculating a definite integral.

arccosine function The inverse of the cosine function, written arccos x or $\cos^{-1}x$. Because the cosine function is periodic, its domain must be restricted in order to have an inverse; the restricted domain is the interval $[0, \pi]$. We thus have the following definition: y = arccos x if and only if x = cos y, where $0 \le y \le \pi$ and $-1 \le x \le 1$. The domain of arccos x is $[-1, 1]$, and its range $[0, \pi]$. Its derivative is $d/dx\ \arccos x = -1/\sqrt{1 - x^2}$.

arcsine function The inverse of the sine function, written arcsin x or $\sin^{-1}x$. Because the sine function is periodic, its domain must be restricted in order to have an inverse; the restricted domain is the interval $[-\pi/2, \pi/2]$. We thus have the following definition: y = arcsin x if and only if x = sin y, where $-\pi/2 \le y \le \pi/2$ and $-1 \le x \le 1$. The domain of arcsin x is $[-1, 1]$, and its range $[-\pi/2, \pi/2]$. Its derivative is $d/dx\ \arcsin x = 1/\sqrt{1 - x^2}$.

arctangent function The inverse of the tangent function, written arctan x or $\tan^{-1}x$. Because the tangent function is periodic, its domain must be restricted in order to have an inverse; the restricted domain is the open interval $(-\pi/2, \pi/2)$. We thus have the following definition: y = arctan x if and only if x = tan y, where $-\pi/2 < y < \pi/2$. The domain of arctan x is all real numbers, that is, $(-\infty, \infty)$; its

Arccosine function

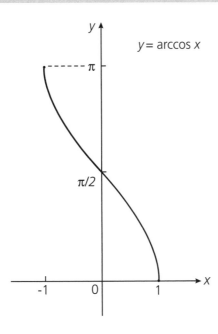

$y = \arccos x$

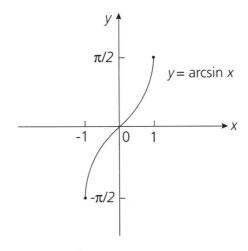

$y = \arcsin x$

Arcsine function

Arctangent function

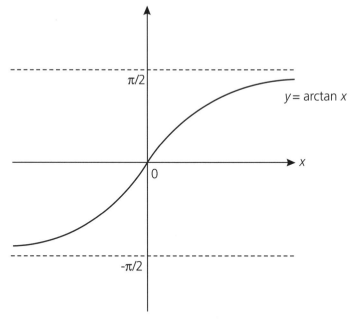

range is $(-\pi/2, \pi/2)$, and the lines $y = \pi/2$ and $y = -\pi/2$ are horizontal asymptotes to its graph. Its derivative is d/dx arctan $x = 1/(1 + x^2)$.

area Loosely speaking, a measure of the amount of two-dimensional space, or surface, bounded by a closed curve. Except for a few simple curves, finding the area involves calculating a definite integral.

area between two curves The definite integral $\int_a^b [f(x) - g(x)]\, dx$, where $f(x)$ and $g(x)$ represent the "upper" and "lower" curves, respectively, and a and b are the lower and upper limits of the interval under consideration.

area function The definite integral $\int_a^x f(t)\, dt$, considered as a function of the upper limit x; that is, we think of $t = a$ as a fixed point and $t = x$ as a variable point, and consider the area under the graph of $y = f(x)$ as a function of x. The letter t is a "dummy variable," used so as not confuse it with the upper limit of integration x.
See also FUNDAMENTAL THEOREM OF CALCULUS.

area in polar coordinates The definite integral $\frac{1}{2}\int_\alpha^\beta [f(\theta)]^2\, d\theta$, where $r = f(\theta)$ is the polar equation of the curve, and α and β are the lower and upper angular limits of the region under consideration.

area of surface of revolution The definite integral $2\pi \int_a^b f(x)\sqrt{1 + [f'(x)]^2}\ dx$, where $y = f(x)$ is the equation of a curve that revolves about the x-axis, and a and b are the lower and upper limits of the interval under consideration. If the graph revolves about the y-axis, we write its equation as $x = g(y)$, and the area is $2\pi \int_c^d g(y)\sqrt{1 + [g'(y)]^2}\ dy$.
See also SOLID OF REVOLUTION.

area under a curve Let $f(x) \geq 0$ on the closed interval [a, b]. The area under the graph of $f(x)$ between $x = a$ and $x = b$ is the definite integral $\int_a^b f(x)\ dx$. If $f(x) \leq 0$ on [a, b], we replace $f(x)$ by $|f(x)|$.

Arithmetic-Geometric Mean Theorem Let $a_1, a_2, \ldots, a_n$ be n positive numbers. The theorem says that $\sqrt[n]{a_1 a_2 \ldots a_n} \leq (a_1 + a_2 + \ldots + a_n)/n$, with equality if, and only if, $a_1 = a_2 = \ldots = a_n$. In words: the geometric mean of n positive numbers is never greater than their arithmetic mean, and the two means are equal if, and only if, the numbers are equal.
See also ARITHMETIC MEAN; GEOMETRIC MEAN.

arithmetic mean of n real numbers $a_1, a_2, \ldots, a_n$ is the expression $(a_1 + a_2 + \ldots + a_n)/n = \dfrac{1}{n}\sum_{i=1}^{n} a_i$. This is also called the *average* of

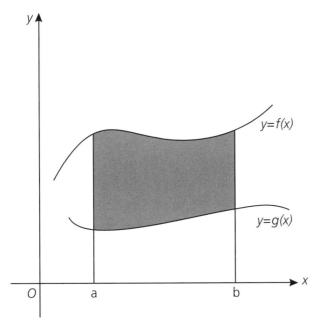

Area between two curves

Area in polar coordinates

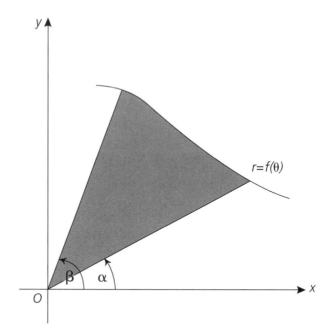

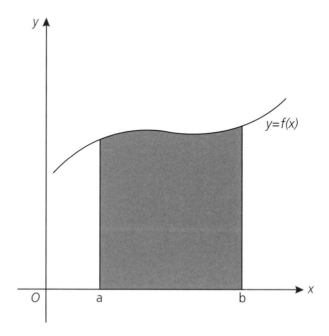

Area under a curve

the n numbers. For example, the arithmetic mean of the numbers
1, 2, –5, and 7 is $(1 + 2 + (-5) + 7)/4 = 5/4 = 1.25$.

asymptote (from the Greek *asymptotus,* not meeting) A straight line to which
the graph of a function $y = f(x)$ gets closer and closer as x approaches
a specific value c on the x-axis, or as x → ∞ or –∞.
 Horizontal: A function has a horizontal asymptote if its graph
approaches the horizontal line $y = c$ as x → ∞ or x → –∞. For
example, the function $y = (2x + 1)/(x - 1)$ has the horizontal
asymptote $y = 2$.
 Slant: A function has a slant asymptote if its graph approaches a
line that is neither horizontal nor vertical. This usually happens
when the degree of the numerator of a rational function is greater
by 1 than the degree of the denominator. For example, the function

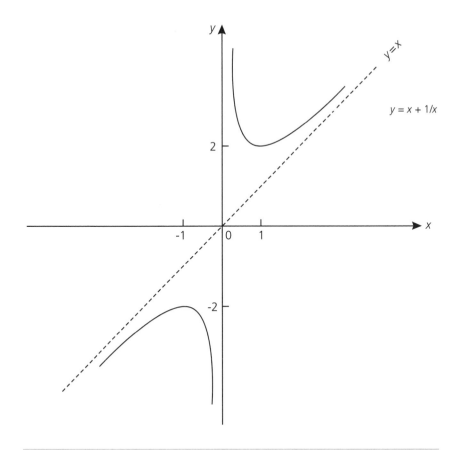

**Slant asymptote of
$y = x + 1/x$**

Asymptotes of
y = (2x + 1)/(x − 1)

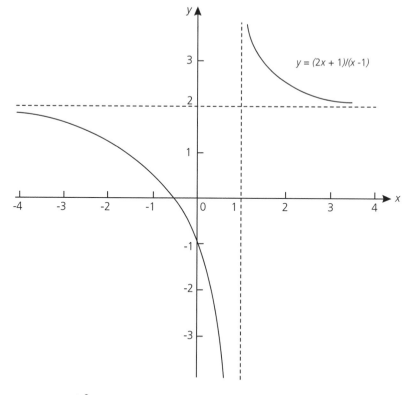

$y = (x^2 + 1)/x = x + 1/x$ has the slant asymptote $y = x$, because as $x \to \pm\infty$, $1/x$ approaches 0.

    Vertical: A function has a vertical asymptote if its graph approaches the vertical line $x = a$ as $x \to a$. For example, the function $y = (2x + 1)/(x − 1)$ has the vertical asymptote $x = 1$.

average Of *n* numbers: Let the numbers be $x_1, x_2, \ldots, x_n$. Their average is the expression $(x_1 + x_2 + \ldots + x_n)/n = \dfrac{1}{n}\sum_{i=1}^{n} x_i$. Also called the *arithmetic mean* of the numbers.

    Of a function: Let the function be $y = f(x)$. Its average over the interval [a, b] is the definite integral $\dfrac{1}{b-a}\int_{a}^{b} f(x)dx$. For example, the average of $y = x^2$ over [1, 2] is $\dfrac{1}{2-1}\int_{1}^{2} x^2 dx = 7/3$.

average cost function A concept in economics. If the cost function of producing and selling *x* units of a commodity is C(x), the average cost per unit is $C(x)/x$, and is itself a function of x. It is measured in dollars per unit.

average rate of change *See* RATE OF CHANGE, AVERAGE.

average velocity Let a particle move along the x-axis. Its position at time t is a function of t, so we write $x = x(t)$ (we are using here the same letter for the dependent variable as for the function itself). The average velocity of the particle over the time interval $[t_1, t_2]$ is the difference quotient $v = \dfrac{x_2 - x_1}{t_2 - t_1}$.

base of logarithms A positive number $b \neq 1$ such that $b^x = y$. We then write $x = \log_b y$.

binomial series The infinite series $(1 + x)^r = 1 + rx + [r(r - 1)/2!]x^2 +$ $[r(r - 1)(r - 2)]/3!]x^3 + \ldots = \sum\limits_{k=0}^{\infty} \binom{r}{k} x^k$, where r is any real number and $-1 < x < 1$. This series is the TAYLOR SERIES for the function $(1 + x)^r$; the symbol $\binom{r}{k} = \dfrac{r!}{k!(r - k)!}$ denotes the binomial coefficients. In the special case when r is a nonnegative integer, the series terminates after $r + 1$ terms and is thus a finite progression.

   See also BINOMIAL THEOREM.

Binomial Theorem The statement that $(a + b)^n = a^n + na^{n-1}b + [n(n - 1)/2!]$ $a^{n-2}b^2 + [n(n - 1)(n - 2)/3!]\, a^{n-3}b^3 + \ldots + nab^{n-1} + b^n$. The kth term $(k = 0, 1, 2, \ldots, n)$ in this expansion is $[n(n - 1)(n - 2) \ldots$ $(n - k + 1)/k!]a^{n-k}b^k$, where $k!$ (read "k factorial") is $1 \cdot 2 \cdot 3 \cdot \ldots \cdot k$ (by definition, $0! = 1$). The coefficients of this expansion are called the *binomial coefficients* and written as $\binom{n}{k}$ or ${}^nC_k$. As an example, $(a + b)^4 = a^4 + 4a^3b + [4(4 - 1)/2!]a^2b^2 + [4(4 - 1)(4 - 2)/3!]ab^3 +$ $[(4(4 - 1)(4 - 2)(4 - 3)/4!]b^4 = a^4 + 4a^3b + 6a^2b^2 + 4ab^3 + b^4$. Note that the expansion is the same whether read from right to left or from left to right.

bounds A number M is an *upper bound* of a sequence of numbers $a_1, a_2, \ldots, a_n$, if $a_i \leq M$ for all i. A number N is a *lower bound* if $a_i \geq N$ for all i. For example, the sequence $1/2, 2/3, 3/4, \ldots, n/(n + 1)$ has an upper bound 1 and a lower bound 0. Of course, any number $M' > M$ is also an upper bound, and any number $N' < N$ is also a lower bound of the same sequence; thus upper and lower bounds are not unique.

Boyle's Law (Boyle-Mariotte Law) A law in physics that relates the pressure P and volume V of a gas in a closed container held at constant temperature. The law says that under these circumstances,

PV = constant; that is if $P_1V_1 = P_2V_2$, where "1" and "2" denote two different states of the gas. Named after the English physicist Robert Boyle (1627–91).

break-even point The number of units x of a commodity that must be produced and sold in order for a business to "break even," that is, to turn loss into profit (in business parlance, to go from "red" to "black"). If $C(x)$, $R(x)$, and $P(x)$ are, respectively, the cost, revenue, and profit functions, we have $P(x) = R(x) – C(x)$. At the break-even point $P(x) = 0$, and so $R(x) = C(x)$. Solving this equation for any given cost and revenue functions gives the desired number x.

calculus, differential *See* DIFFERENTIAL CALCULUS.

calculus, integral *See* INTEGRAL CALCULUS.

cardioid A heart-shaped curve whose polar equation is $r = 1 + \cos \theta$. It has a cusp at $(0, 0)$ pointing to the right (the equation $r = 1 + \sin \theta$ describes a similar cardioid with a cusp at $(0, 0)$ pointing up). The cardioid is a special case of the *Limaçon,* whose polar equation is $r = b + a \cos \theta$.

Cartesian coordinates (rectangular coordinates) In the plane, an ordered pair of numbers (x, y), where x is the distance of a point P from the

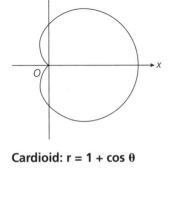

Cardioid: $r = 1 + \cos \theta$

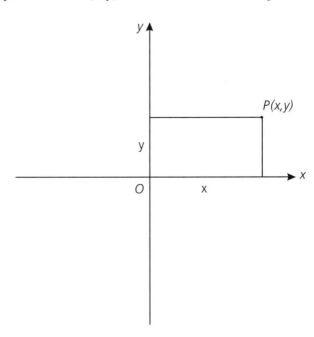

Cartesian coordinates in two dimensions

Cartesian coordinates in three dimensions

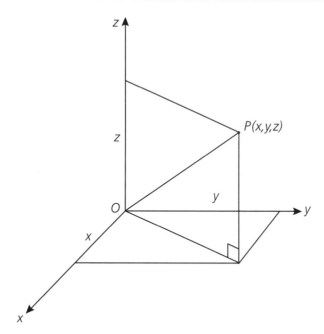

y-axis, and y is its distance from the x-axis. In space, an ordered triplet of numbers (x, y, z). They are named after their inventor, RENÉ DESCARTES.

catenary From the Latin word *catena* (chain), a curve whose equation is $y = a \cosh x/a = a(e^{x/a} + e^{-x/a})/2$, where a is constant. A chain hanging freely under the force of gravity has the shape of a catenary.

Cauchy-Schwarz Inequality The inequality $|a_1b_1 + a_2b_2 + \ldots + a_nb_n|^2 \leq (|a_1|^2 + |a_2|^2 + \ldots + |a_n|^2)(|b_1|^2 + |b_2|^2 + \ldots + |b_n|^2)$ for any real numbers $a_1, \ldots, a_n$ and $b_1, \ldots, b_n$. Equality holds if, and only if, $a_1/b_1 = a_2/b_2 = \ldots = a_n/b_n$. Named after AUGUSTIN-LOUIS CAUCHY and the German mathematician Hermann Amandus Schwarz (1843–1921).

For integrals: The inequality $|\int_a^b f(x)g(x)\ dx|^2 \leq (\int_a^b |f(x)|^2\ dx) (\int_a^b |g(x)|^2\ dx)$.

Equality holds if, and only if, $f(x)/g(x) = $ constant.

center of mass (center of gravity, centroid) The point at which a physical system must be balanced in order to maintain its equilibrium under

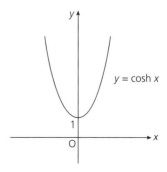

Catenary

the force of gravity. For a one-dimensional discrete system of n particles with masses m_i and positions x_i, i = 1, 2, . . ., n, the center of mass is given by the formula $X = \left(\sum_{i=1}^{n} m_i x_i\right)\Big/\left(\sum_{i=1}^{n} m_i\right)$.

For a two-dimensional system with masses at (x_i, y_i), the center of mass has coordinates (X, Y), where $X = \left(\sum_{i=1}^{n} m_i x_i\right)\Big/\left(\sum_{i=1}^{n} m_i\right)$ and $Y = \left(\sum_{i=1}^{n} m_i y_i\right)\Big/\left(\sum_{i=1}^{n} m_i\right)$.

Analogous formulas exist for a three-dimensional system. In the case of a one-dimensional *continuous* system with a density function $\rho(x)$, the center of mass is given by the formula $X = \left(\int_a^b x\rho(x)dx\right)\Big/\left(\int_a^b \rho(x)dx\right)$, the limits of integration being determined by the physical dimensions of the system. Similar formulas exist for two- and three-dimensional continuous systems, but they involve double and triple integrals.

centroid Center of mass of a solid with constant density. For example, the centroid of a triangle of uniform thickness is the intersection of its medians.
See also CENTER OF MASS.

Chain Rule If y = f(u) and u = g(x), then the derivative of the *composite function* y = f(g(x)) = h(x) is given by h′(x) = f′(g(x))g′(x) = f′(u)g′(x); in Leibniz's "d" notation, this is equivalent to dy/dx = (dy/du)(du/dx), where u = g(x) is the "inner function" and y = f(u) the "outer function." The expression g′(x) = du/dx is the "inner derivative." For example, if y = $(3x + 2)^5$, we write y = u^5 where u = 3x + 2; then y′ = (dy/du)(du/dx) = $(5u^4)(3) = 15u^4 = 15(3x + 2)^4$ (the last step is necessary because we want to write the answer in terms of *x,* not *u*). The rule can be extended to any number of "component" functions; thus if y = f(g(h(x))), then dy/dx = (dy/du)(du/dv)(dv/dx) = f′(u)g′(v)h′(x), where v = h(x), u = g(h(x)) = g(v), and y = f(g(h(x))) = f(u) (hence the name "chain rule").

change of base The base *a* in the *exponential function* y = a^x can be changed to the *natural base e* by using the formula $a^x = e^{(\ln a)x}$. The base *a* in the *logarithmic function* y = $\log_a x$ can be changed to any other base *b* by using the formula $\log_a x = (\log_b x)/(\log_b a)$. For example, $\log_2 x = (\log_{10} x)/(\log_{10} 2) \approx (\log_{10} x)/0.30103$. The most common change of base is from base 10 *(common logarithms)* to base *e* *(natural logarithms):* $\log_{10} x = (\ln x)/(\ln 10) \approx (\ln x)/2.30259$; here "ln" means natural logarithm.

change of variable *See* SUBSTITUTION, METHOD OF.

chaos   A modern branch of mathematics dealing with phenomena in which a small change in the parameters can lead to a large change in the outcome. This has been popularized by the saying, "a butterfly flapping its wings in California may trigger an earthquake in Japan." Chaos is most efficiently studied by computer simulation, rather than by seeking exact solutions of the equations governing the phenomenon under consideration. One example is weather patterns, which can be dramatically affected by a small change in local circumstances such as temperature, pressure, and humidity.

characteristic equation   Consider the linear, homogeneous differential equation with constant coefficients $a_n y^{(n)} + a_{n-1} y^{(n-1)} + \ldots + a_1 y' + a_0 y = 0$, where $y = f(x)$ and $y^{(i)}$, $i = 1, \ldots, n$ denotes the ith derivative of y with respect to x. The substitution $y = c e^{rx}$, where c and r are as yet undetermined constants, transforms this equation into the *algebraic* equation $a_n r^n + a_{n-1} r^{n-1} + \ldots + a_0 = 0$ (note that the expression ce^{rx} cancels in the process). This equation is the *characteristic equation* associated with the given differential equation; it is a polynomial of degree n in the unknown r. By solving it for r, we find the possible solutions of the differential equation, whose linear combination gives us the *general* solution. For example, the differential equation $y'' + 5y' + 6y = 0$ has the characteristic equation $r^2 + 5r + 6 = (r + 2)(r + 3) = 0$, whose roots are $r = -2$ and $r = -3$. Thus the equation has the two solutions $y_1 = Ce^{-2x}$ and $y_2 = De^{-3x}$. The general solution is formed by a linear combination of these two solutions: $y = Ce^{-2x} + De^{-3x}$. The coefficients C and D are arbitrary coefficients; they can only be determined from the *initial conditions* associated with the differential equation.

   If the roots of the characteristic equation are *complex conjugates,* then their imaginary part can be rewritten as a linear combination of sine and cosine functions. For example, the differential equation $y'' + 2y' + 4y = 0$ has the characteristic equation $r^2 + 2r + 4 = 0$, whose roots are $r = -1 + i\sqrt{3}$ and $r = -1 - i\sqrt{3}$. Thus the differential equation has the general solution $y = Ce^{(-1 + i\sqrt{3})x} + De^{(-1 - i\sqrt{3})x}$. This is equivalent to the expression $y = e^{-x} (A \cos\sqrt{3}x + B \sin\sqrt{3}x)$, signifying damped oscillations.

   If the characteristic equation has repeated roots, for example a double solution r, then the solution of the differential equation is a linear combination of the functions e^{rx} and xe^{rx}. For example, the differential equation $y'' - 4y + 4 = 0$ has the characteristic equation $r^2 - 4r + 4 = 0$, which has the double root $r = 2$. The general solution of the differential equation is $y = Ae^{2x} + Bxe^{2x} = (A + Bx)e^{2x}$.

   See also DIFFERENTIAL EQUATION; LINEAR COMBINATION.

circle, general equation of The equation $Ax^2 + Ay^2 + Bx + Cy + D = 0$ represents a circle; depending on the values of the coefficients, this circle can be *real, imaginary,* or *degenerate* (a single point). Examples follow:

   The equation $x^2 + y^2 - 10x + 6y + 18 = 0$ represents a real circle with center at $(5, -3)$ and radius 4.
   The equation $x^2 + y^2 - 10x + 6y + 38 = 0$ represents an imaginary circle with center at $(5, -3)$ and "radius" 2i.
   The equation $x^2 + y^2 - 10x + 6y + 34 = 0$ represents a degenerate circle (the point $(5, -3)$).

   To change the general equation of a circle to the *standard equation,* we need to complete the squares on *x* and *y*.
   See also CIRCLE, STANDARD EQUATION OF.

circle, standard equation of The equation $(x - h)^2 + (y - k)^2 = r^2$ represents a circle of radius *r* and center at the point (h, k). For example, the equation $(x - 5)^2 + (y + 3)^2 = 16$ represents a circle with radius 4 and center at $(5, -3)$. If $h = k = 0$ and $r = 1$, we get the equation of the *unit circle.*

Clairaut equation The differential equation $y = xy' + f(y')$, where *f* is a given function of y'. Named after ALEXIS-CLAUDE CLAIRAUT.

closed interval *See* INTERVAL.

coefficient A constant multiplying the variable part in an algebraic expression. For example, the coefficient of $-7xy^2$ is -7 (however, if *y* is held constant, the coefficient of the same expression is $-7x$; if *x* is held constant, the coefficient is $-7y^2$). The coefficient of $3\cos 2x$ is 3, since $\cos 2x$ is regarded as the variable part.

common logarithm *See* LOGARITHM, COMMON.

comparison tests for improper integrals Let *f* and *g* be continuous functions with $0 \le f(x) \le g(x)$ for all $x \ge a$. Then:

(1). If $\int_a^\infty g(x)\, dx$ is convergent, so is $\int_a^\infty f(x)\, dx$.

(2). If $\int_a^\infty f(x)\, dx$ is divergent, so is $\int_a^\infty g(x)\, dx$.

For example, on the interval $(1, \infty)$, $e^{-x^2} \le e^{-x}$ and therefore $\int_1^\infty e^{-x^2} dx \le \int_1^\infty e^{-x} dx$; since the second integral converges to $1/e$, the first integral will also converge, though not to the same limit.

comparison tests for proper integrals (1). Let $f(x) \ge g(x)$ for all *x* in the interval [a, b]. Then $\int_a^b f(x)\, dx \ge \int_a^b g(x)\, dx$. In particular, if $f(x) \ge 0$ on [a, b], then $\int_a^b f(x)\, dx \ge 0$.

(2). Let $m \leq f(x) \leq M$ for all x on [a, b]. Then $m(b - a) \leq \int_a^b f(x)\,dx$ $\leq M(b - a)$. For example, on the interval [1, 2], $1 \leq \sqrt{x} \leq 2$; therefore $1 \cdot (2 - 1) \leq \int_1^2 \sqrt{x}\,dx \leq 2 \cdot (2 - 1)$, so the value of $\int_1^2 \sqrt{x}\,dx$ is between 1 and 2 (it is, in fact, about 1.21895).

comparison test for series Let $\sum_{i=1}^{\infty} a_i$ and $\sum_{i=1}^{\infty} b_i$ be two series with positive terms and $a_i \leq b_i$ for all i. Then:

(1). If $\sum_{i=1}^{\infty} b_i$ converges, so does $\sum_{i=1}^{\infty} a_i$.

(2). If $\sum_{i=1}^{\infty} a_i$ diverges, so does $\sum_{i=1}^{\infty} b_i$.

complex conjugates The conjugate of the complex number $a + ib$ is the complex number $a - ib$; for example, the conjugate of $5 + 7i$ is $5 - 7i$, and vice versa. The conjugate of the imaginary number $3i$ is the imaginary number $-3i$; the conjugate of the real number 2 is 2, because either can be written as $2 + 0i$.
See also COMPLEX NUMBER.

complex number A number of the form $a + ib$, where a and b are real numbers and $i^2 = -1$ (or equivalently, $i = \sqrt{-1}$). A complex number is often denoted by a single letter, usually z; we write $z = a + ib$, where $a = $ Re z (read: "the real part of z") and $b = $ Im z ("the imaginary part of z"). If $b = 0$, the number is *real*; if $a = 0$, it is *imaginary*. Thus the set of real numbers (and also the set of imaginary numbers) is a subset of the set of complex numbers.
See also POLAR FORM OF A COMPLEX NUMBER.

composite function A combination of two or more functions so that the output of one function is the input to the other. Symbolically, if $y = f(u)$ and $u = g(x)$, then $y = f(g(x))$ is the *composition* of g and f (in that order). For example, the function $y = \sqrt{1 + x}$ can be regarded as a composition of the functions $u = g(x) = 1 + x$ and $y = f(u) = \sqrt{u}$. Generally $f(g(x))$ is different from $g(f(x))$; in the example just given, $g(f(x)) = g(\sqrt{x}) = 1 + \sqrt{x}$, which is different from $\sqrt{1 + x}$. Sometimes the symbol $(f \circ g)(x)$ is used for $f(g(x))$.

compound interest A financial procedure whereby a bank pays interest not only on the money invested (the principal), but also on the *interest* accumulated from the investment. Put differently, at the end of each compounding period the bank takes the *current* balance and regards it as if it had just been reinvested at the same interest rate. If the principal is denoted by P, the annual interest rate by r, and the money is compounded n times a year, then the balance A after t years is given by the formula $A = P(1 + r/n)^{nt}$. [Note: when using this formula, always change r to a decimal.]

For example, if P = $100, r = 5% = 0.05, and n = 12 (monthly compounding), then the balance after 10 years will be $A = 100(1 + 0.05/12)^{120} = \$164.70$.

See also SIMPLE INTEREST.

Continuous: If the bank compounds the investment *continuously* (that is, every instant) at the annual interest rate r (also called the *nominal interest rate*), the balance A after t years is given by the formula $A = Pe^{rt}$, where e is the base of natural logarithm. In the example given above, the balance after 10 years will be $A = 100e^{0.5} = \$164.87$.

See also FUTURE VALUE; PRESENT VALUE.

concavity A measure of the bending of a curve. A curve is *concave up* at a point $x = x_0$ if it lies above the tangent line to the curve at x_0 (more precisely, at all points in an open interval around x_0). A curve is *concave down* at a point if it lies below the tangent line to the curve at x_0 (more precisely, at all points in an open interval around x_0). Concavity is related to the *second derivative* of the function representing the curve.

See also CONCAVITY TEST; INFLECTION POINT.

concavity test Let $y = f(x)$ be a twice-differentiable function at a point $x = x_0$. If $f''(x_0) > 0$, the graph of f is *concave up* at x_0. If $f''(x_0) < 0$, the

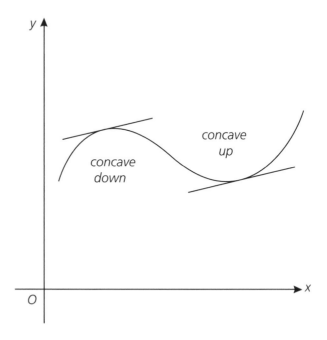

Concavity

   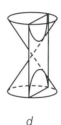 

a   *b*   *c*   *d*   *e*

Conic sections: (a) circle;
(b) ellipse; (c) parabola;
(d) hyperbola; (e) pair
of lines

graph of f is *concave down* at x_0. If $f''(x_0) = 0$, the graph may be concave up or concave down at x_0, or it may be flat there; the test in this case is inconclusive.

As examples, consider the functions $f(x) = x^2$, $g(x) = x^3$, and $h(x) = x^4$ at $x_0 = 0$. We have $f''(x) = 2 > 0$, so the graph of f (a parabola) is concave up at 0 (indeed on the entire *x*-axis). On the other hand $g''(x) = 6x$ and $h''(x) = 12x^2$, so both $g''(x)$ and $h''(x)$ are 0 at 0. Yet the graph of h (a parabola-like shape) is concave up at 0, while that of g is flat there (it changes from concave down to concave up at 0).

See also CONCAVITY; INFLECTION POINT.

conditional convergence *See* CONVERGENCE, CONDITIONAL.

conic sections If a cone is sliced by a plane, the cross section is a *conic section*. If the plane does not pass through the cone's vertex, the conic section is a *circle,* an *ellipse,* a *parabola,* or a *hyperbola,* depending on the angle of inclination of the plane to the cone's axis. If the plane passes through the vertex, we get a pair of straight lines, which may be regarded as a limiting case of a hyperbola (a "degenerate hyperbola").

conjugate, complex *See* COMPLEX CONJUGATES.

constant function The function $y = f(x) = c$, where c is a constant. Its graph is a horizontal line with *Y*-intercept at $(0, c)$.

constant of integration An arbitrary constant that is added to an indefinite integral, or antiderivative. For example, the antiderivative of x^2 is $x^3/3 + C$. *See also* ANTIDERIVATIVE; INTEGRAL, INDEFINITE.

continuity Intuitively speaking, a function is continuous if its graph does not have any breaks; that is, if we can draw it with one stroke of the pen. More precisely, a function $y = f(x)$ is continuous if a small change in x results in a small change in y. This can be stated mathematically as follows: $f(x)$ is continuous at a point $x = a$ if it is defined there and if $\lim_{x \to a} f(x) = f(a)$. All polynomial functions are

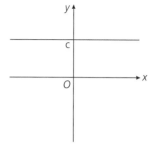

Constant function y = c

continuous everywhere, as are the functions sin x, cos x, and e^x. A rational function is continuous except for those x values for which the denominator is zero.

See also DISCONTINUITY.

continuous compounding *See* COMPOUND INTEREST, CONTINUOUS.

convergence Absolute: An infinite series $\sum_{i=1}^{\infty} a_i = a_1 + a_2 + \ldots$ is said to be *absolutely convergent* if the series $\sum_{i=1}^{\infty} |a_i|$ (that is, all terms of the original series being replaced by their absolute values) converges. For example, the series $1 - 1/2 + 1/4 - 1/8 + - \ldots$ is absolutely convergent, because the series $1 + 1/2 + 1/4 + 1/8 + \ldots$ is convergent (the former converges to 2/3, the latter to 2). Note that if $\sum_{i=1}^{\infty} |a_i|$ converges, so does $\sum_{i=1}^{\infty} a_i$, and if $\sum_{i=1}^{\infty} a_i$ *diverges,* so does $\sum_{i=1}^{\infty} |a_i|$, but the converse of these statements is false.

Conditional: An infinite series $\sum_{i=1}^{\infty} a_i = a_1 + a_2 + \ldots$ is said to be *conditionally convergent* if it converges but the series $\sum_{i=1}^{\infty} |a_i|$ diverges. For example, the series $1 - 1/2 + 1/3 - 1/4 + - \ldots$ is conditionally convergent, because it converges (its sum is ln 2), but the series $1 + 1/2 + 1/3 + 1/4 + \ldots$ (the harmonic series) diverges.

Of an improper integral: The integral $\int_a^{\infty} f(x)\, dx$ is said to be convergent if $\lim_{b \to \infty} \int_a^b f(x)\, dx$ exists (i.e., is a finite number). For example, $\int_1^b 1/x^2\, dx$ converges to the limit 1 as $b \to \infty$; we write $\int_1^{\infty} 1/x^2\, dx = 1$.

Of a sequence: A sequence is said to converge if its terms approach a limit as the number of terms increases beyond bound; in symbols, the sequence $a_1, a_2, a_3, \ldots$ converges to the limit L if $\lim_{i \to \infty} a_i = L$; we also write $a_i \to L$ as $i \to \infty$. For example, the sequence 1/1, 1/2, 1/3, . . . converges to the limit 0 as $i \to \infty$. A formal definition is as follows: the sequence $a_1, a_2, a_3, \ldots$ converges to the limit L if for every positive number ε, no matter how small, we can find a corresponding number N such that $|a_i - L| < \varepsilon$ whenever $i > N$; that is, we can make the difference (in absolute value) between the terms of the sequence and its limit as small as we please by going sufficiently far out in the sequence. In the example given, if we want the terms 1/i to be closer to 0 than, say, 1/1,000, we can do this by letting i be greater than 1,000; that is, $|1/i - 0| < 1/1,000$ whenever $i > 1,000$.

Of a series: A series is said to converge if its *sequence of partial sums* converges to a limit *S;* in symbols, the series $\sum_{i=1}^{\infty} a_i = a_1 + a_2 + a_3 + \ldots$ converges to the sum *S* if the sequence $a_1, (a_1 + a_2), (a_1 + a_2 + a_3)$, ... converges to the limit *S* as $n \to \infty$. We write $\lim_{n \to \infty} \sum_{i=1}^{n} a_i = S$, or briefly, $\sum_{i=1}^{\infty} a_i = S$. For example, the geometric series $1 + 1/2 + 1/4 + 1/8 + \ldots$ converges to the limit 2 as the number of terms increases beyond bound, so we write $\sum_{i=1}^{\infty} \frac{1}{2^{i-1}} = 2$.

Radius of: *See* POWER SERIES.

convergence tests *See* Section Four D.

coordinates Rectangular: *See* RECTANGULAR COORDINATES.
Polar: *See* POLAR COORDINATES.

cosecant function The function $y = f(x) = 1/\sin x$, denoted by csc x. Its domain is all real numbers except $x = 0, \pm\pi, \pm 2\pi, \ldots$ (these are the

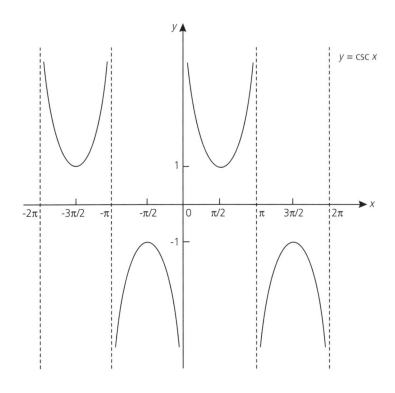

$y = \csc x$

Cosecant function

values for which sin x = 0). Its range is the compound interval $(-\infty, -1] \cup [1, \infty)$. The graph of csc x has vertical asymptotes at $x = 0, \pm\pi, \pm2\pi, \ldots$ and is periodic with period 2π. The derivative of the cosecant function is $d/dx \ \csc x = -\cos x/\sin^2 x = -\cot x \ \csc x$.

See Section Four A for other properties of the cosecant function.

cosine function   The function $y = f(x) = \cos x$. Its domain is all real numbers, and its range the interval $[-1, 1]$. Its graph is periodic—it repeats every 2π radians. In applications, particularly in vibration and wave phenomena, the vertical distance from the *x*-axis to either the highest or the lowest point of the graph (that is, 1) is called the *amplitude,* and the period 2π is the *wavelength.* More generally, the function $y = a \cos bx$ has amplitude $|a|$ and period $2\pi/b$. One can also shift the graph to the left or right; this is represented by the function $y = a \cos (bx + c)$. The derivative of the cosine function is $d/dx \ \cos x = -\sin x$. For other properties of the cosine function, see Section Four A.

See also SINE FUNCTION.

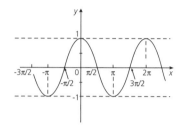

Cosine function y = cos x

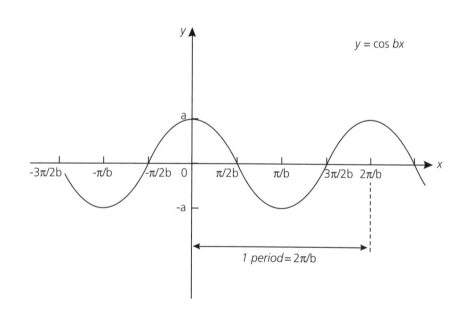

Cosine function
y = a cosb x

cotangent function The function $y = 1/\tan x = \cos x/\sin x$, denoted by $\cot x$ (sometimes $\operatorname{ctg} x$). Its domain is all real numbers except $0, \pm\pi, \pm2\pi, \pm3\pi, \ldots$ (these are the values for which $\sin x = 0$), at which the graph of $\cot x$ has vertical asymptotes. Its range is all real numbers. The cotangent function is periodic with period π. The derivative of the cotangent function is $d/dx \cot x = -1/\sin^2 x = -\csc^2 x$.

    See also Section Four A for additional properties of the cotangent function; TANGENT FUNCTION.

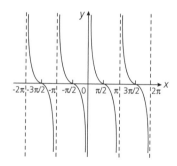

Cotangent function

critical number (value, point) A value of x for which the derivative $f'(x)$ of a function is either zero or undefined. For example, the critical numbers of $f(x) = 2x^3 + 3x^2 - 36x + 4$ are $x = -3$ and 2, because $f'(x) = 6x^2 + 6x - 36 = 6(x-2)(x+3) = 0$ has the solutions $x = -3$ and 2. The critical number of $f(x) = x^{2/3}$ is $x = 0$, because $f'(x) = (2/3)x^{-1/3} = 2/(3x^{1/3})$ is undefined at $x = 0$.

cubic function A polynomial of degree 3 with real coefficients; that is, the function $f(x) = ax^3 + bx^2 + cx + d$, where a, b, c, and d are constants and $a \neq 0$. The graph of a cubic function has at most one maximum point and one minimum point (it may have neither), and it always has one inflection point.

curvature A measure of the amount of bending of a graph. Curvature is expressed mathematically by the formula $\kappa = |y''|/[1 + (y')^2]^{3/2}$, where $y = f(x)$ is the equation of the graph (κ is the Greek letter "kappa"). The quantity $\rho = 1/\kappa$ is called the *radius of curvature* (ρ is the Greek letter "rho"); it is generally a function of x and varies from point to point (except for a circle, in which case ρ is the radius of the circle).

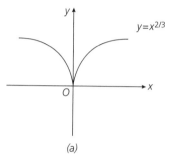

(a)

curve Loosely speaking, "curve" is synonymous with "graph." More precisely, a curve is a set of ordered pairs (x, y) in which x and y are related by an equation, or in which each is a function of a third variable t (a parameter). A curve can exist in two dimensions (a planar curve), or in three dimensions (a spatial curve); in the latter case, it is a set of ordered triples (x, y, z) in which $x, y,$ and z are each a function of a parameter t.

cusp A point where a curve has a corner, that is, where it makes an abrupt change in direction. At a cusp, the derivative does not exist, while the tangent line may or may not exist. For example, the functions $y = x^{2/3}$ and $y = |x|$ both have a cusp at $x = 0$; the first function has a vertical tangent at $x = 0$, while the second has no tangent line there.

    See also PARAMETRIC EQUATIONS.

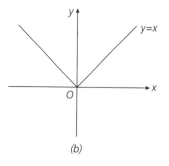

(b)

Cusp: (a) vertical tangent line at 0; (b) no tangent line at 0

cycloid   A curve traced by a point on the circumference of a circle as it rolls along a straight line. If the straight line is the x-axis and the circle has radius a, the parametric equations of the cycloid are $x = a(\theta - \sin \theta)$, $y = a(1 - \cos \theta)$. The distance between two adjacent cusps is equal to the circumference $2\pi a$ of the circle. An inverted cycloid is the curve along which an object will slide down under the force of gravity in the shortest possible time.
   See also INTRODUCTION.

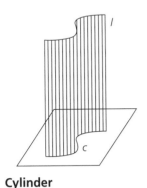

Cylinder

cylinder   In the narrow sense, the surface of a solid in the shape of a soft-drink can (a right circular cylinder). More generally, the surface generated when a straight line l in space moves parallel to itself while intersecting a planar curve c. l is called the *generator* and c the *generating curve.*

decibel   A unit of loudness. A sound of intensity I (in watts/cm^2) has a decibel loudness $dB = 10 \log I/I_o$, where "log" stands for common (base 10) logarithm, and I_o is the *threshold intensity* (the lowest sound intensity the ear can still perceive). Because it is a logarithmic scale, the decibel scale compresses an enormous range of intensities into a relatively narrow range of loudness levels. For example, the loudness level of a quiet conversation is about 50 dB, while that of a loud rock concert can be as high as 120 dB. Every doubling of the intensity increases the loudness level by 10 log 2, or about 3 dB.
   See also RICHTER SCALE.

degree of a differential equation   The highest power of the highest-order derivative of the unknown function y appearing in the equation. For example, the equation $xy' + y^2 = \ln x$ is of degree 1.

demand function   In economics, a function $p = f(x)$ that gives the price consumers are willing to pay for each unit of a commodity, when x units are being produced and sold. Sometimes the inverse $x = g(p)$ of this function is being used.

dependent variable   The variable y in the function $y = f(x)$. Its value depends on our choice of x (the *independent variable*), hence the name.

derivative   The value of $\lim\limits_{h \to 0} [f(x + h) - f(x)]/h$; that is, the limit of the difference quotient of a function $y = f(x)$ at a given point x in its domain, as the increment h tends to zero (provided this limit exists). The derivative is denoted by $f'(x)$, or simply by y'. To indicate that the derivative is being evaluated at a specific point $x = a$, we write $f'(a)$ or $y'|_{x=a}$.

An alternative notation, due to Leibniz, is dy/dx, or d/dx f(x); when evaluated at the point x = a, we write $(dy/dx)_{x=a}$. Because *x* can be any number at which this limit exists, the derivative itself is a function of *x;* this is manifest in the notation f′(x).

As an example, consider the function y = f(x) = x^2. Its derivative is $\lim_{h \to 0}$ [(x + h)^2 – x^2]/h. Of course, we cannot simply substitute h = 0 in this limit, because this will give us the indeterminate expression 0/0. We go around this by first simplifying the expression inside the limit: [(x + h)^2 – x^2]/h = [x^2 + 2xh + h^2 – x^2]/h = (2xh + h^2)/h = h(2x + h)/h = 2x + h. Now we let h → 0, resulting in the expression 2x. Thus f′(x) = 2x.

The derivative can be interpreted in two ways: as the *rate of change* of the independent variable *y* with respect to the dependent variable *x,* or as the *slope of the tangent line* to the graph of y = f(x) at the point *x.*

The concept of derivative is the cornerstone of the *differential calculus.* There are several rules that allow us to find derivatives in shorter ways than actually finding the limit; these are known as the rules of differentiation, and they form the backbone of the calculus course.

See also DIFFERENCE QUOTIENT; DIFFERENTIATION, RULES OF.

difference quotient The ratio [f(x + h) – f(x)]/h, where f(x) is a given function. The numerical value of this ratio is the slope of the secant line to the graph of y = f(x) through the points P(x, f(x)) and Q(x + h, f(x + h)). Also called the *rise-to-run ratio,* or the *average rate of change,* and often denoted by Δy/Δx, where Δx and Δy are the increments in *x* and *y,* respectively. As an example, for the function f(x) = x^2 we have [f(x + h) – f(x)]/h = [(x + h)^2 – x^2]/h; after expanding the expression (x + h)^2 and simplifying, this becomes 2x + h.

differentiable function A function that has a derivative at a given point in its domain; that is, a function for which $\lim_{h \to 0}$ [f(x + h) – f(x)]/h exists.

For example, y = x^2 is differentiable everywhere (that is, for all *x*), while y = |x| is differentiable for all x except x = 0.

See also DERIVATIVE.

differential Loosely speaking, an "infinitely small" change in a variable. If the variable is *x,* its differential is written dx. The derivative dy/dx of a function y = f(x) can be interpreted as the ratio of the two differentials dy and dx. Thus, instead of writing f′(x) = dy/dx, we can "cross multiply" and write dy = f′(x) dx, which is convenient when

Difference quotient

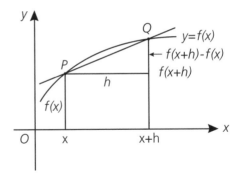

changing the variable in an integral. We should point out that some mathematicians reject this interpretation and insist on regarding the derivative as the limit of the difference quotient $\Delta y / \Delta x$ as $\Delta x \to 0$. Nevertheless, applied scientists routinely think of Δx as if it were a differential dx and set up their equations accordingly.

See also SUBSTITUTION, METHOD OF.

differential calculus The part of calculus that deals with rates of change, or derivatives, of functions.

differential equation An equation involving an unknown function $y = f(x)$ and its derivatives. For example, the equation $xy' + y = 0$ is a differential equation of *first order* (meaning that the highest derivative is the first); its solution is the function $y = c/x$, where c is an arbitrary constant; this is called the *general solution,* because of our freedom to choose any value for c.

If, in addition, we impose on y an *initial condition,* then we can determine c. In the example just given, suppose we require that the graph of the unknown function should pass through the point $(1, 2)$; substituting $x = 1$ and $y = 2$ into the general solution $y = c/x$, we get $c = 2$, giving us the *particular solution* $y = 2/x$.

As another example, consider the equation $y'' + y = 0$, which is of *second order* because it involves the second derivative of the unknown function *y;* its general solution is $y = A \cos x + B \sin x$, where A and B are two arbitrary constants (generally, the number of arbitrary constants is equal to the order of the equation).

The study of differential equations is an important branch of higher mathematics, with numerous applications in every field of science.

differential operator A symbol whose meaning is "take the derivative of . . .". It is usually denoted by d/dx or *D*. Anything standing to its immediate right is to be differentiated. For example, $d/dx(x^2) = 2x$; this can also be written $Dx^2 = 2x$. For a second differentiation, we write d^2/dx^2 or D^2; thus $D^2x^2 = D(D(x^2)) = D(2x) = 2$. A differential operation has the advantage of compactness; it can often be treated according to simple algebraic rules, making it convenient to work with when solving differential equations.

differentiation The act of finding the derivative of a function.
See also DIFFERENTIATION, RULES OF; DERIVATIVE; Section Four B.

differentiation formulas See Section Four B.

differentiation, rules of

$d/dx[cf(x)] = c\, d/dx\, f(x)$, where c = constant
$d/dx\, [f(x) \pm g(x)] = d/dx\, f(x) \pm d/dx\, g(x)$ (Sum Rule)
$d/dx\, [f(x)g(x)] = f(x)\, d/dx\, g(x) + g(x)\, d/dx\, f(x)$ (Product Rule)
$d/dx\, [f(x)/g(x)] = [g(x)\, d/dx\, f(x) - f(x)\, d/dx\, g(x)]/[g(x)]^2$, provided
$\quad g(x) \neq 0$ (Quotient Rule)
These rules are often written in abbreviated form: $(cf)' = cf'$,
$(f \pm g)' = f' \pm g'$, $(fg)' = fg' + gf'$, and $(f/g)' = (gf' - fg')/g^2$.
See also POWER RULE; PRODUCT RULE; QUOTIENT RULE; and SUM RULE for examples of each rule.

directrix See PARABOLA.

discontinuity Loosely speaking, any "break" in the graph of a function; more precisely, a situation where an arbitrarily small increment of the independent variable leads to a large increment of the dependent variable. A function $f(x)$ can have a discontinuity at $x = a$ if any of the following three conditions happens:

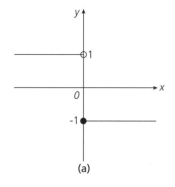

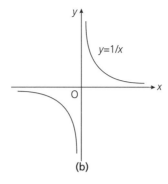

Discontinuity: (a) finite; (b) infinite

(1). f(a) does not exist.

(2). $\lim_{x \to a}$ f(x) does not exist.

(3). f(a) and $\lim_{x \to a}$ f(x) both exist, but $\lim_{x \to a}$ f(x) ≠ f(a).

Discontinuities can be of three types:

Finite, as in the function f(x) = {1 when x < 0 and –1 when x ≥ 0} at the point x = 0. This corresponds to case (2) above.

Infinite, as in the function f(x) = 1/x at x = 0. This corresponds to case (1) above.

Removable, as in the function f(x) = (sin x)/x. This function is undefined at x = 0, so its graph has a "hole" there. We can assign an arbitrary value to f(x) at 0 (for example f(0) = 2), but this would not fill the hole, and the function would still be discontinuous at x = 0 (case (3) above). But if we define f(0) to be 1, this will fill the hole and make the function continuous, because $\lim_{x \to 0}$ (sin x)/x = 1.

See also CONTINUITY.

discrete mathematics (finite mathematics) The branch of mathematics that deals with noncontinuous processes where the limit concept does not play a role. An example is graph theory, the study of the connectiveness of a discrete system of points and lines.

See also ANALYSIS.

disk method A method of finding the volume of a solid of revolution by imagining it to be sliced into infinitely many thin parallel disks centered on the axis of revolution. The volume is then found by integrating the volumes of these disks over the length of the solid. If the solid is generated by revolving the graph of y = f(x) about the *x*-axis, the required volume is given by $V = \pi \int_{a}^{b} [f(x)]^2 \, dx$, where *a* and *b* are the endpoints of the interval in question. As an example, a circular cone of base-radius *r* and height *h* can be generated by revolving the line y = (r/h)x about the *x*-axis. The volume is $V = \pi \int_{0}^{h} [(r/h)x]^2 \, dx = \pi r^2 h/3$.

See also SHELL METHOD; SOLID OF REVOLUTION.

distance formula Between two points x_1 and x_2 on the *x*-axis: d = |$x_2 - x_1$|. For example, the distance between the points $x_1 = 3$ and $x_2 = -5$ is |(–5) – 3| = |–8| = 8.

Between two points (x_1, y_1) and (x_2, y_2) in the plane:

$$d = \sqrt{(x_2 - x_1)^2 + (y_2 - y_1)^2}.$$

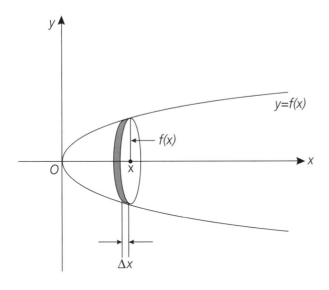

For example, the distance between (2, –3) and (1, 5) is
$d = \sqrt{[(1-2)^2 + (5-(-3))^2]} = \sqrt{65}$.

Between a point (x_0, y_0) and a line $Ax + By = C$ in the plane:

$$d = |Ax_0 + By_0 - C|/\sqrt{A^2 + B^2}.$$

For example, the distance between the point (2, –3) and the line
$4x + 5y = 1$ is

$$d = |4 \cdot 2 + 5 \cdot (-3) - 1|/\sqrt{4^2 + 5^2} = 8/\sqrt{41}.$$

divergence Of an improper integral: The improper integral $\int_a^\infty f(x)\,dx$ is said
to be divergent if $\lim_{b\to\infty} \int_a^b f(x)\,dx$ does not exist. For example,
$\lim_{b\to\infty} \int_1^b 1/x\,dx$ does not exist, so the integral diverges.
    Of a sequence: A sequence is said to *diverge* if it does not have a
limit as the number of terms increases beyond bound. For example, the
sequence of natural numbers, 1, 2, 3, . . ., n, . . . does not have a limit as
$n \to \infty$ and thus diverges. For a sequence to diverge, its terms do not
necessarily have to get larger and larger: the sequence 1, –1, 1, –1, . . .
alternates between 1 and –1 but does not have a limit, so it diverges.
    Of a series: A series is said to diverge if the sequence of its partial
sums does not have a limit. For example, the series $1 + 1/2 + 1/3 + \ldots$

+ 1/n + . . . (the harmonic series) does not approach a limit as n $\to \infty$ (even though its terms get smaller and smaller with *n*), so it diverges.

See also CONVERGENCE; IMPROPER INTEGRAL; POWER SERIES; SERIES, SEQUENCE OF PARTIAL SUMS OF.

domain The set of elements (usually numbers) that can be used as an input to a function. For example, the domain of f(x) = 1/(x + 3) consists of all real numbers except –3, that is, the set $\{x|\ x \neq 3\}$. The domain of g(x) = 1/$\sqrt{(x + 3)}$ is the set $\{x|\ x > -3\}$ (assuming that we consider only real numbers as outputs of the function).

dummy variable The summation index in a sum when using the sigma notation, or the variable of integration in a definite integral. The word *dummy* comes from the fact that we can change the letter for the index or variable of integration without affecting the outcome. Thus, it makes no difference if we write $\sum_{i=1}^{n} a_i$, or $\sum_{j=1}^{n} a_j$, or $\sum_{k=1}^{n} a_k$, because the subscript simply plays the role of a counter. Similarly, it makes no difference if we write $\int_{a}^{b} f(x)\ dx$, or $\int_{a}^{b} f(y)\ dy$, or $\int_{a}^{b} f(z)\ dz$, since the outcome is a number. Note that this is not so for an *indefinite* integral: $\int f(x)\ dx$ is a function of *x,* while $\int f(y)\ dy$ is a function of *y.*

e (base of natural logarithms) The limit of $(1 + 1/n)^n$ as n $\to \infty$, and the sum of the infinite series 1 + 1/1! + 1/2! + 1/3! + ... Its approximate value is 2.7182818284. Like π, *e* is an irrational number, so its decimal expansion is nonterminating and nonrepeating (it is also a *transcendental* number). Its importance in calculus comes from the fact that the exponential function with base *e,* y = e^x, is equal to its own derivative. The inverse of this function, $\log_e x$, is called the *natural logarithm* of *x* and written ln x.

See also EXPONENTIAL FUNCTION; LOGARITHMIC FUNCTION; TRANSCENDENTAL NUMBER.

eccentricity Of the ellipse $x^2/a^2 + y^2/b^2 = 1$ (where a > b): the ratio e = c/a, where c = $\sqrt{a^2 - b^2}$ is the distance from the center of the ellipse to either of its two foci. For example, the eccentricity of the ellipse $x^2/25 + y^2/9 = 1$ is e = $\sqrt{25 - 9}/5$ = 4/5 = 0.8. The eccentricity of an ellipse is always less than 1; the smaller the eccentricity, the more "circular" the ellipse is.

See also ELLIPSE.

Of the hyperbola $x^2/a^2 - y^2/b^2 = 1$: the ratio e = c/a, where c = $\sqrt{a^2 + b^2}$ is the distance from the center of the hyperbola to either of its two foci. For example, the eccentricity of the hyperbola

$x^2/25 - y^2/9 = 1$ is $e = (\sqrt{25 + 9})/5 = \sqrt{34}/5 \approx 1.166$. The eccentricity of a hyperbola is always greater than 1.

See also HYPERBOLA.

elasticity of demand The expression $E(x) = (p/x)/(dp/dx)$, were x is the number of units of a commodity being demanded, and p is the price per unit, regarded as a function of x (the *demand function*).

elementary functions The family of functions consisting of polynomials and ratios of polynomials (that is, rational functions), power functions and their inverses (radicals), trigonometric and exponential functions and their inverses (the latter include logarithms), and any finite combination of these functions.

ellipse The set of points (x, y) in the plane, the sum of whose distances from two fixed points is constant. The two fixed points are the *foci* (single: focus) of the ellipse, and the point midway between them is the *center*. If the foci are on the *x*-axis at $(-c, 0)$ and $(c, 0)$, then the center is at the origin $(0, 0)$ and the axes of the ellipse are along the coordinate axes, with the long axis along the *x*-axis. In this case, if we denote the sum of the distances from any point on the ellipse to its two foci by

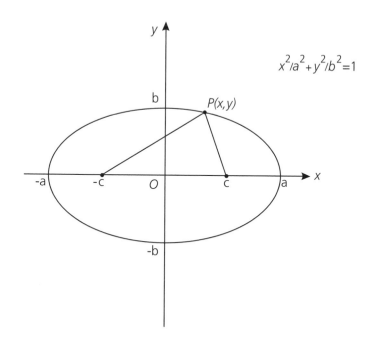

$$x^2/a^2 + y^2/b^2 = 1$$

Ellipse

2a, the equation of the ellipse is $x^2/a^2 + y^2/b^2 = 1$, where $b = \sqrt{a^2 - c^2}$. The points (a, 0) and (–a, 0) are the *vertices* (single: vertex) of the ellipse. The line segment joining the two vertices is the *major axis;* its length is 2a. The line segment through the center and perpendicular to the major axis is the *minor axis;* its length is 2b. If the foci are on the *y*-axis at (0, –c) and (0, c), the equation of the ellipse is $x^2/b^2 + y^2/a^2 = 1$, with the major axis along the *y*-axis.

See also ECCENTRICITY.

Latus rectum of: A line segment through either focus of the ellipse and perpendicular to its major axis; its length is $2b^2/a$.

Reflective property of: If we imagine the ellipse to be coated with a reflective material, then a ray of light emanating from one focus will be reflected to the other focus, regardless of which direction the ray is aimed at.

epsilon-delta (ε-δ) *See* LIMIT.

error A term having various meanings, depending on the subject under discussion. The most common uses are:

Absolute error: The difference between the exact value of a quantity and an approximation of it (e.g., the value measured in an experiment, or the value given up to a prescribed number of decimal places). In symbols, if x_0 denotes the exact value and x the approximate value, then the error is $x_0 - x$ (in some applications the error is defined as $|x_0 - x|$). The error (also known as *true error*) is usually denoted by ε (Greek epsilon).

Relative error: the ratio of the absolute error to the exact value: $(x_0 - x)/x_0$. This ratio is often given as a percent. For example, if the "true" value of π is taken as 3.1415927 (the value returned by an eight-digit handheld calculator), and if this value is approximated by 3.14, then the absolute error is $3.1415927 - 3.14 = 0.0015927$, and the relative error is $0.0015927/3.1415927 = 0.000507$, or about 0.05 percent. In calculus, one often uses differentials to estimate an error.

See also DIFFERENTIALS; LINEAR APPROXIMATION.

even function A function f with the property $f(-x) = f(x)$ for all x values in its domain. Examples are $f(x) = x^2$ (and in fact x raised to any even power, hence the name), cos x, and $1/(1 - x^2)$. Geometrically, the graph of an even function is symmetric about the *y*-axis.

See also ODD FUNCTION; SYMMETRY.

exponential decay A term applied to a variable y that decreases (often with time) according to the formula $y = y_0 e^{-kt}$, where y_0 is the initial value of y (the value when t = 0), and k is a positive constant (but note the

Exponential decay

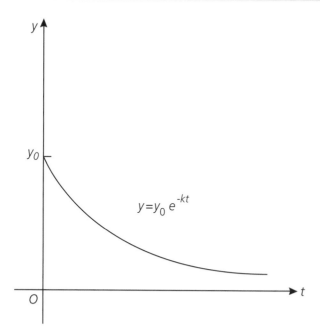

minus sign in front of it). Examples are radioactive decay and the depreciation of the monetary value of a commodity. Often the term applies to variations of the above formula, such as $y = a + y_0 e^{-kt}$, where a is a constant.

See also EXPONENTIAL GROWTH; HALF-LIFE.

exponential function The function $y = f(x) = b^x$, where the base b is a positive number different from 1.

Natural: The exponential function with base $e \approx 2.7182818$, that is, $y = e^x$. It has the property that $y' = y$; that is, the function is equal to its derivative. Often the term *natural exponential function* is used for the more general function $y = ce^{kx}$, where c and k are constants; c is the *initial value* (the value of y when $x = 0$), and k measures the *rate of growth* of the function.

Power series of: The infinite series $1 + x + x^2/2! + x^3/3! + \ldots$, which converges to e^x for all values of x.

Properties of: Let $f(x) = b^x$. Then $f(0) = 1$, $f(x_1 + x_2) = f(x_1) \cdot f(x_2)$, and $f'(x) = (\ln b) b^x$, where ln denotes natural logarithm (logarithm base e).

See also LOGARITHMIC FUNCTION.

**Exponential function:
(a) growing; (b) decaying**

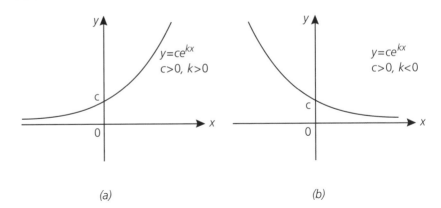

(a)   (b)

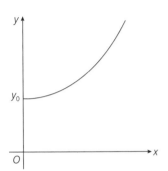

Exponential growth

exponential growth A term applied to a variable y that increases (often with time) according to the formula $y = y_0 e^{kt}$, where y_0 is the initial value of y (the value when $t = 0$), and k is a positive constant. For example, the growth of a population with time is approximately exponential, from which the popular phrase "exponential growth" comes. Often the term is applied to variations of the above formula, such as $y = a + y_0 e^{kt}$, where a is a constant.
 See also EXPONENTIAL DECAY; EXPONENTIAL FUNCTION.

extreme value of a function The smallest or largest value of a function on an interval.

Extreme Value Theorem If a function f is continuous on the closed interval [a, b], it has at least one minimum value and one maximum value on that interval. This is an example of an *existence theorem;* it does not tell us *how* to find these extreme values, only that they exist. The requirement that f be continuous on the interval is essential; also, the theorem may not apply if the interval is open at one or both endpoints.

extremum (plural: *extrema*) A maximum or minimum.

factorial The product $1 \cdot 2 \cdot 3 \cdot \ldots \cdot n$, denoted by n!. The first ten factorials are $1! = 1$, $2! = 2$, $3! = 6$, $4! = 24$, $5! = 120$, $6! = 720$, $7! = 5,040$, $8! = 40,320$, $9! = 362,880$, and $10! = 3,628,800$. We also define $0! = 1$. Factorials often appear in expressions involving permutations and combinations, as also in many infinite series.

family of curves A set of curves represented by a common equation; the individual curves are obtained by assigning different values to a

Extreme Value Theorem

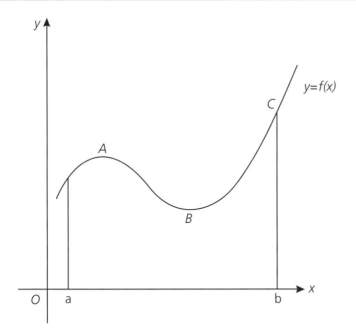

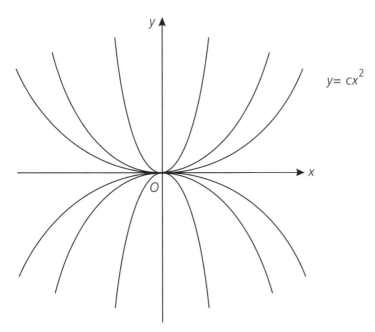

Family of curves: $y = cx^2$

constant, or *parameter,* that appears in the equation. For example, the equation $y = cx^2$ represents the family of parabolas with vertex at the origin and axis along the y-axis. The individual curves of a family share common properties that can be deduced from the equation of the family.

first derivative *See* DERIVATIVE.

First Derivative Test Let c be a critical number of the function $y = f(x)$ on an open interval (a, b); that is, the derivative $f'(x)$ is either 0 or undefined at $x = c$. Then, if f' changes sign from positive to negative at c, f has a relative maximum at c; if f' changes sign from negative to positive at c, f has a relative minimum at c; and if f' does not change sign at c, f has neither a relative maximum nor a relative minimum at c. Examples are $y = -x^2$, $y = x^{2/3}$, and $y = x^3$, respectively.
    See also CRITICAL NUMBER; MAXIMUM, RELATIVE; MINIMUM, RELATIVE.

first-order differential equation A differential equation in which the highest derivative of the unknown function is the first derivative: an example is $xy' + y^2 = \sin x$.

focus   Of an ellipse, *See* ELLIPSE.
     Of a hyperbola, *See* HYPERBOLA.
     Of a parabola, *See* PARABOLA.

Fourier series An infinite series of sine and cosine terms of the form

$$a_0/2 + \sum_{n=1}^{\infty} (a_n \cos nx + b_n \sin nx)$$
$$= a_0/2 + a_1 \cos x + b_1 \sin x + a_2 \cos 2x + b_2 \sin 2x + \ldots$$

Under certain conditions, a function $f(x)$, regarded as a periodic function over the interval $[-\pi, \pi]$, can be represented by a Fourier series. The coefficients are found from the following formulas:

First Derivative Test: (a) f **has a relative maximum at x = 0; (b)** f **has a relative minimum at x = 0; (c)** f **has neither a maximum nor a minimum at x = 0**

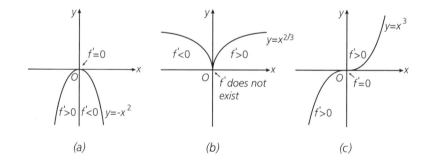

(a)   (b)   (c)

Fourier series: (a) the function f(x) = x, f(x + 2π) = f(x); (b) the first four partial sums of the Fourier series of f(x)

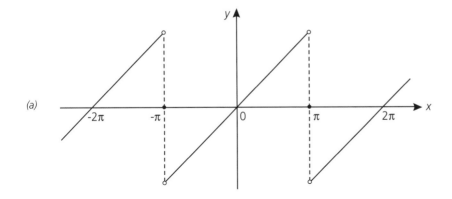

(a)

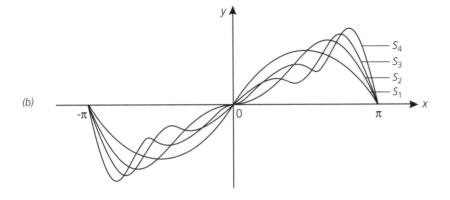

(b)

$$a_n = (1/\pi) \int_{-\pi}^{\pi} f(x) \cos nx\, dx, \quad n = 0, 1, 2, \ldots$$
$$b_n = (1/\pi) \int_{-\pi}^{\pi} f(x) \sin nx\, dx, \quad n = 1, 2, \ldots$$

(note that the first formula applies also for n = 0, giving us the coefficient a_0).

For example, the function f(x) = x, regarded as a periodic function over [–π, π], is represented by the series 2[(sin x)/1 – (sin 2x)/2 + (sin 3x)/3 – + . . .], which has only sine terms. Fourier series are used in physics to describe vibration and wave phenomena. The series is named after its discoverer, JEAN-BAPTISTE-JOSEPH FOURIER.

function(s) Algebraic: *See* ALGEBRAIC FUNCTIONS.

Average value of: The average value of a function y = f(x) over an interval [a, b] is defined as $\dfrac{1}{b-a} \int_a^b f(x)\, dx$. For example, the

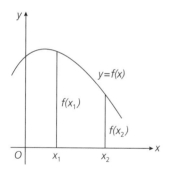

$y = f(x)$

$f(x_1)$

$f(x_2)$

Decreasing function

average value of $y = x^2$ over the interval $[1, 2]$ is $\frac{1}{2-1}\int_1^2 x^2 dx = 7/3$.

Geometrically, the average value of $f(x)$ is the height of the rectangle with base $[a, b]$ whose area is equal to the area under the graph of $f(x)$ from $x = a$ to $x = b$.

Composite: *See* COMPOSITE FUNCTION.

Constant: *See* CONSTANT FUNCTION.

Continuous: *See* CONTINUITY.

Cubic: *See* CUBIC FUNCTION.

Decreasing: A function $f(x)$ is decreasing *on a interval* if, given any two numbers x_1 and x_2 in the interval, $f(x_1) > f(x_2)$ whenever $x_1 < x_2$. A function $f(x)$ is decreasing *at a point c* in its domain if its derivative is negative at $x = c$; that is, if $f'(c) < 0$.

(a)
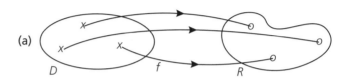

(b)
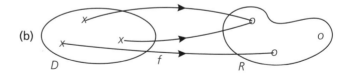

(c)
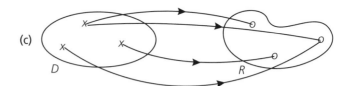

(d)
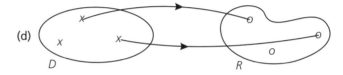

Function, definition of: (a) a function; (b) a function; (c) not a function; (d) not a function

Definition of: A function f is a rule of correspondence, or "mapping," between two sets D and R, such that every element of D corresponds to ("is mapped on") exactly one element of R. The set D is called the *domain* of f, and the set R the *range*. We can think of the function as a set of *ordered pairs* $\{(x, y)\}$ in which each x is an element of the domain and each y an element of the range, and no two y correspond to the same x. We also require that *every* element of D must appear as the first element in one pair (in other words, that no element of D is "left out").

For example, if $D = \{1, 2, 3\}$ and $R = \{3, 7, \pi\}$, the set of ordered pairs $\{(1, 3), (2, 7), (3, \pi)\}$ is a function, as is the set $\{(1, 3), (2, 7), (3, 7)\}$ (the same y can come from two different x), but the set $\{(1, 3), (2, 7), (1, \pi)\}$ is not; neither is the set $\{(1, 3), (2, 7)\}$, because not all elements of D are being used. The rule of correspondence can be entirely arbitrary, as in the example just given, or it can be an empirical rule (for example, the set of hourly temperature readings at some location over a 24-hour period). In calculus, however, a function is usually given by a *formula* that tells us how to obtain each y from x. For example, the formula $y = 2x + 1$ tells us to take any number x, double it, and then add 1.

See also DOMAIN; FUNCTION NOTATION; ONTO FUNCTION; RANGE.

Derivative of, *See* DERIVATIVE.

Domain of, *See* DOMAIN.

Elementary, *See* ELEMENTARY FUNCTIONS.

Even, *See* EVEN FUNCTION.

Exponential, *See* EXPONENTIAL FUNCTION.

Extreme value of, *See* EXTREME VALUE OF A FUNCTION.

Graph of: The set of all points (x, y) for which y is a given function of x.

Greatest integer, *See* GREATEST INTEGER FUNCTION.

Hyperbolic, *See* HYPERBOLIC FUNCTIONS.

Increasing: A function $f(x)$ is increasing on a interval if, given any two numbers x_1 and x_2 in the interval, $f(x_1) < f(x_2)$ whenever $x_1 < x_2$. A function $f(x)$ is increasing *at a point c* in its domain if its derivative is positive at $x = c$; that is, if $f'(c) > 0$.

Integrable: A function is integrable over an interval $[a, b]$ if the definite integral $\int_a^b f(x)\,dx$ exists (i.e., is a real number). Any function that is continuous on $[a, b]$ is integrable there.

Inverse, *See* INVERSE FUNCTION.

Limit of, *See* LIMIT OF A FUNCTION.

Linear, *See* LINEAR FUNCTION.

Logarithmic, *See* LOGARITHMIC FUNCTION.

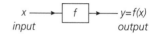

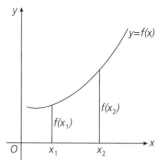

Definition of a function

Increasing function

Maximum on an interval: The largest value a function can attain on that interval.

Minimum on an interval: The smallest value a function can attain on that interval.

Notation: the symbol y = f(x) (read: "*y* is a function of *x*"). This is usually followed by a formula that expresses *y* explicitly as a function of *x*. For example, the formula y = f(x) = 2x + 1 tells us to take a number from the domain (in this case, all real numbers), double it, and add 1. Thus we have f(3) = 2 · (3) + 1 = 7, f(0) = 2 · (0) + 1 = 1, f(a) = 2a + 1, f(2a) = 2(2a) + 1 = 4a + 1, f(x + 1) = 2(x + 1) + 1 = 2x + 3, and so on.

Another way of looking at the functional notation is f( ) = 2( ) + 1, where the blank space is to be filled with the desired value or expression, taken from the domain of *f*. (Note: the parentheses in f(x) do *not* indicate multiplication.) We can think of *x* as an "input" to the function, and of *y* as the corresponding "output." The function itself is denoted by *f* (of course, any other letter would do, such as *g* or *h*).

Odd, *See* ODD FUNCTION.

One-to-one, *See* ONE-TO-ONE FUNCTION.

Onto, *See* ONTO FUNCTION.

Periodic, *See* PERIODIC FUNCTION.

Polynomial, *See* POLYNOMIAL FUNCTION.

Quadratic, *See* QUADRATIC FUNCTION.

Range of, *See* RANGE.

Rational, *See* RATIONAL FUNCTION.

Transcendental, *See* TRANSCENDENTAL FUNCTION.

Trigonometric, *See* TRIGONOMETRIC FUNCTIONS.

Zero of, *See* ZERO OF A FUNCTION.

Fundamental Theorem of Calculus First form: Let F(x) be an antiderivative of f(x), that is, $F'(x) = f(x)$. Then $\int_a^b f(x)\,dx = F(b) - F(a)$

The last expression is also written as $F(x)\big|_a^b$.

Second form:

$$\frac{d}{dx} \int_a^x f(t)\,dt = f(x)$$

Here we regard the integral as a function of its upper limit *x*, so we use *t* for the variable of integration to distinguish it from *x*.

See also ANTIDERIVATIVE; AREA FUNCTION.

future value The amount of money, or balance, in a bank account at the end of a specified time period since the money was deposited. If the amount deposited (the principal) is *P*, the annual interest rate is *r* (expressed as a decimal), and the money is compounded n times a year for t years, then the future value A is given by the formula $A = P(1 + r/n)^{nt}$.

For example if $P = \$100$, $r = 5\%$, and $n = 12$ (that is, the money is compounded monthly), then the future value after $t = 5$ years is $A = 100(1 + 0.05/12)^{12 \cdot 5} = \$128.34$. If the bank is using *continuous* compounding, the formula is $A = Pe^{rt}$, where $e \sim 2.78$ is the base of natural logarithms. For the data given above, the future value for continuous compounding will be $A = 100e^{0.05 \cdot 5} = \$128.40$.

See also PRESENT VALUE.

generalized harmonic series *See* p-SERIES.

generalized power rule *See* POWER RULE, GENERALIZED.

general solution of a differential equation A solution that contains arbitrary constants. For example, the general solution of the equation $y'' + y = 0$ is $y = A \cos x + B \sin x$, where A and B are arbitrary constants.

See also PARTICULAR SOLUTION OF A DIFFERENTIAL EQUATION.

geometric mean Of n positive numbers $a_1, a_2, \ldots, a_n$ is the expression $\sqrt[n]{(a_1 a_2 \ldots a_n)}$. For example, the geometric mean of the numbers 1, 2, 3, and 5 is $\sqrt[4]{(1 \cdot 2 \cdot 3 \cdot 5)} = \sqrt[4]{30} \approx 2.34$.

geometric progression A progression, or sequence, of numbers in which each number is obtained from its predecessor by multiplication by a constant number, called the *quotient* of the progression. Examples are $1, 2, 4, 8, 16, \ldots, 2^{n-1}, \ldots$ (here the initial term is 1 and the quotient is 2), $1, 1/2, 1/4, 1/8, 1/16, \ldots, 1/2^{n-1}, \ldots$ (initial term 1, quotient 1/2), and $1, -1, 1, -1, \ldots, (-1)^{n-1}, \ldots$ (initial term 1, quotient -1). Generally, if the initial term is a and the quotient q, we can write the progression as $a, aq, aq^2, \ldots, aq^{n-1}, \ldots$, where aq^{n-1} is the nth term (note that the first term can be written as aq^0, which explains the $n - 1$ in the nth term). A geometric progression can be finite or infinite. If finite, it ends after the term aq^{n-1}, in which case we drop the three dots behind it.

geometric series The sum of the terms of a geometric progression. For example, $1 + 2 + 4 + 8 + \ldots + 2^{n-1}$. A geometric series can be finite or infinite. If finite, we can write it as $a + aq + aq^2 + \ldots + aq^{n-1}$, where a is the initial term, q the quotient, and n the number of terms (note that the first term can be written as aq^0, which explains the $n - 1$ in the nth term). In this case the sum of the series can be found from the formula $S = a(1 - q^n)/(1 - q)$. For the example given above, the sum is $S = 1 \cdot (1 - 2^n)/(1 - 2) = 2^n - 1$. If the series is infinite and if $|q| < 1$ (that is, q is between -1 and 1), then the series converges to the sum $S = a/(1 - q)$. For example, the infinite geometric series $1 + 1/2 + 1/4 + 1/8 + \ldots + 1/2^{n-1} + \ldots$ has the sum $S = 1/[1 - (1/2)] = 2$. If $|q| \geq 1$, the series diverges, and its sum is undefined.

See also CONVERGENCE, DIVERGENCE, GEOMETRIC PROGRESSION.

graph   Of an equation: the set of points (x, y) whose x and y coordinates satisfy a given equation in x and y. For example, the graph of the equation $x^2 + y^2 = 1$ is the unit circle (the circle with center at $(0, 0)$ and radius 1).

Of a function: the set of points (x, y), where y is a given function of x. For example, the graph of the function $y = \sqrt{1 - x^2}$ is the upper half of the unit circle; the lower half is the graph of $y = -\sqrt{1 - x^2}$.

greatest integer function   The function $y = [\![x]\!]$ defined as follows: for any $x, [\![x]\!]$ is the greatest integer not exceeding x. Thus $[\![2.1]\!] = 2$, $[\![2.9]\!] = 2$, $[\![2.999]\!] = 2$, but $[\![3]\!] = 3$. Also, $[\![0]\!] = 0$, $[\![-2.1]\!] = -3$, and so on. In practical terms, $[\![x]\!]$ is obtained by rounding x *downward* to the nearest integer, unless x itself is an integer, in which case $[\![x]\!] = x$. The graph of $y = [\![x]\!]$ has a staircase-like shape, with a jump of 1 at each integer value of x.

growth and decay   *See* EXPONENTIAL DECAY; EXPONENTIAL GROWTH.

half-life   The time it takes a radioactive substance to decay to one half of its original mass. Different substances have vastly different half-life times; for example, the half-life of the ordinary isotope of uranium (U^{238}) is about 4,510,000,000 years; that of ordinary radium (Ra^{226}) is 1,600 years, and that of Ra^{220} just 23 milliseconds. The term "half-life" can be applied to any quantity that decays exponentially with time. If the quantity decays according to the formula $y = y_0 e^{-kt}$ (where y_0 is the initial quantity and k a positive constant), then the half-life, denoted by the Greek letter τ (tau) is given by $\tau = (\ln 2)/k$.

See also EXPONENTIAL DECAY.

half-open interval   An interval that is open at one endpoint and closed at the other. If the open endpoint is on the left and the closed endpoint on the right, we denote the interval by $(a, b]$; if the closed endpoint is on the left and the open endpoint on the right, by $[a, b)$.

harmonic series   The series whose terms are the reciprocals of the natural numbers: $1 + 1/2 + 1/3 + 1/4 + 1/5 + \ldots$. This series diverges, though extremely slowly. The name "harmonic" comes from the association of this series with the harmonics of a vibrating string.

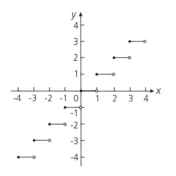

Greatest integer function

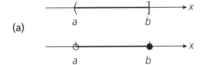

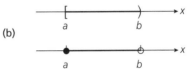

Half-open interval showing two notations: (a) open-closed; (b) closed-open

higher-order derivative A *second-order* derivative, or *second derivative* for short, is the derivative of the derivative of a function $y = f(x)$; it is denoted by d^2y/dx^2 (which comes from the expression $d/dx(dy/dx)$), or by $f''(x)$, or, if it is clear which variable is to be differentiated, simply by y''. A *third-order* derivative is the derivative of the second derivative, denoted by d^3y/dx^3, $f'''(x)$, or y'''. For example, if $y = 3x^4$, then $y' = 12x^3$, $y'' = 36x^2$, and $y''' = 72x$. In physics, the first derivative of the position function of a moving object is its *velocity,* and the second derivative is its *acceleration.*

horizontal asymptote *See* ASYMPTOTE, HORIZONTAL.

horizontal line A line parallel to the *x*-axis. Its equation is $y = b$, where *b* is a constant, and its *Y*-intercept is (0, b). For example, the equation $y = -3$ represents a horizontal line through the point (0, –3). A horizontal line has slope zero; consequently, its slope-intercept equation is $y = 0x + b$.

Horizontal Line Test A test that allows us to see if a function $y = f(x)$ whose graph is given has an *inverse:* f has an inverse if and only if every horizontal line intersects its graph at one point at most. For example, the function $y = f(x) = x^2$ (a parabola) does not have an inverse, because the horizontal line $y = 9$ intersects the graph at two points, (3, 9) and (–3, 9).

　　See also INVERSE FUNCTION; ONE-TO-ONE FUNCTION; VERTICAL LINE TEST.

horizontal shift of a graph Let the graph have the equation $y = f(x)$. If we shift (translate) it *c* units to the right (where *c* is a positive number), its new equation is $y = f(x - c)$. If shifted to the left *c* units (where *c* is still positive), its equation becomes $y = f(x + c)$. For example, the graph of $y = (x + 1)^2$ is identical to the graph of $y = x^2$ (a parabola) but shifted one unit to the left.

　　See also VERTICAL SHIFT OF A GRAPH.

hyperbola The set of points (x, y) in the plane for which the difference of the distances from any point to two fixed points is constant. The two

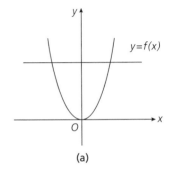

(a)

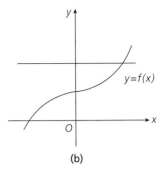

(b)

Horizontal line test: (a) function does not have an inverse; (b) function has an inverse

Hyperbola

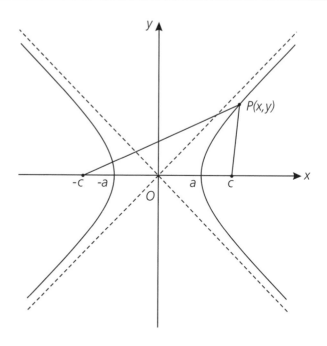

fixed points are the *foci* (single: *focus*) of the hyperbola, and the point midway between them is the *center*. If the foci are at $(-c, 0)$ and $(c, 0)$, then the center is at the origin $(0, 0)$ and the hyperbola is lined up with the coordinate axes. In this case, if we denote the difference of the distances from any point on the hyperbola to its two foci by $2a$, the equation of the hyperbola is $x^2/a^2 - y^2/b^2 = 1$, where $b = \sqrt{c^2 - a^2}$. The points $(a, 0)$ and $(-a, 0)$ are the *vertices* (single: *vertex*) of the hyperbola. The line segment joining the two vertices is the *transverse axis;* its length is $2a$. The line segment of length $2b$ through the center and perpendicular to the transverse axis is the *conjugate axis.* The hyperbola just described has its transverse axis along the x-axis (that is, horizontal). If the foci are at $(0, -c)$ and $(0, c)$, then the transverse axis is along the y-axis (vertical), and the equation of the hyperbola is $y^2/a^2 - x^2/b^2 = 1$.

The graph of every hyperbola consists of two disconnected branches, so we cannot draw it with one stroke of the hand. Every hyperbola is associated with a pair of straight lines that pass through the center and "point the way to infinity." These lines are the *asymptotes* of the hyperbola. For the hyperbola $x^2/a^2 - y^2/b^2 = 1$, the asymptotes have the equations $y = \pm(b/a)x$; for the hyperbola $x^2/b^2 - y^2/a^2 = 1$, the equations are $y = \pm(a/b)x$. Note that the asymptotes themselves are not part of the graph of a hyperbola, but they make it easier to draw the graph.

The shape of a hyperbola is determined by the ratio $e = c/a$, called its *eccentricity*. For any hyperbola, the eccentricity is always greater than 1 (compare with the ellipse). A small eccentricity (that is, e close to 1) corresponds to a hyperbola with narrow branches, while a large value of e corresponds to the branches becoming more open.

hyperbolic functions A class of six functions defined as follows:

$$\sinh x = (e^x - e^{-x})/2$$
$$\cosh x = (e^x + e^{-x})/2$$
$$\tanh x = (e^x - e^{-x})/(e^x + e^{-x})$$
$$\operatorname{csch} x = 2/(e^x - e^{-x})$$
$$\operatorname{sech} x = 2/(e^x + e^{-x})$$
$$\coth x = (e^x + e^{-x})/(e^x - e^{-x})$$

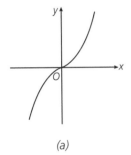

(a)

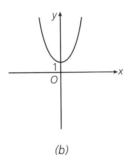

(b)

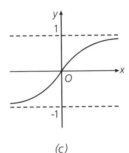

(c)

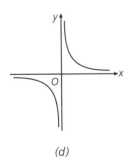

(d)

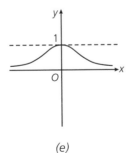

(e)

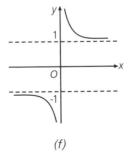

(f)

Hyperbolic functions: (a) $y = \sinh x$; (b) $y = \cosh x$; (c) $y = \tanh x$; (d) $y = \operatorname{csch} x$; (e) $y = \operatorname{sech} x$; (f) $y = \coth x$

(read "hyperbolic sine," etc.), where e^x is the natural exponential function. The domain D and range R of these functions are as follows:

$$
\begin{array}{lll}
\sinh x: & D = (-\infty, \infty) & R = (-\infty, \infty) \\
\cosh x: & D = (-\infty, \infty) & R = [1, \infty) \\
\tanh x: & D = (-\infty, \infty) & R = (-1, 1) \\
\operatorname{csch} x: & D = (-\infty, 0) \cup (0, \infty) & R = (-\infty, 0) \cup (0, \infty) \\
\operatorname{sech} x: & D = (-\infty, \infty) & R = (0, 1] \\
\coth x: & D = (-\infty, 0) \cup (0, \infty) & R = (-\infty, -1) \cup (1, \infty)
\end{array}
$$

The Gateway Arch, St. Louis, Mo. The arch has the shape of an inverted hyperbolic cosine graph. (Courtesy of U.S. Department of the Interior, National Park Service Photo)

The hyperbolic functions exhibit many analogies with the trigonometric functions sin x, cos x, tan x, csc x, sec x, and cot x. For example, tanh x = sinh x/cosh x, sinh (−x) = −sinh x, cosh (−x) = cosh x, $\cosh^2 x - \sinh^2 x = 1$, sinh (x ± y) = sinh x cosh y ± cosh x sinh y, cosh (x ± y) = cosh x cosh y ± sinh x sinh y, and d(sinh x)/dx = cosh x, d(cosh x)/dx = sinh x. Generally, every relation among the trigonometric functions has its hyperbolic counterpart with, however, a possible change in sign in one term (as in the identity $\cosh^2 x - \sinh^2 x = 1$). With appropriate restrictions on the domains of cosh x and sech x, all six hyperbolic functions have inverses, denoted by $\sinh^{-1}$ x, etc. The hyperbolic functions are useful in expressing certain indefinite integrals in a convenient way. The shape of a uniform chain hanging freely under the force of gravity is that of the graph of y = cosh x.

identities, trigonometric *See* Section Four A.

identity function The function y = f(x) = x.

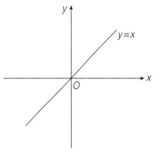

Identity function

implicit differentiation A procedure that allows one to find the derivative of a function, even if this function is not given explicitly in the form y = f(x). For example, the equation $x^2 + y^2 = 1$ defines *two* explicit functions of y in terms of x, namely $y = +\sqrt{1 - x^2}$ and $y = -\sqrt{1 - x^2}$ (the upper and lower half of the unit circle, respectively). Still, we can find the derivatives of these functions without solving the equation $x^2 + y^2 = 1$ for y in terms of x. We differentiate both sides of the equation with respect to x, bearing in mind that y is an unknown function of x. We get 2x + 2yy′ = 0, where the y′ in the second term comes from the chain rule (that is, $d(y^2)/dx = [d(y^2)/dy][dy/dx] = 2yy'$). Solving this last equation for y′, we get y′ = −x/y, which gives us the derivative y′ in terms of x *and y.* (As an option, we can replace y by either $+\sqrt{1 - x^2}$ or $-\sqrt{1 - x^2}$ and obtain y′ as a function of x alone.)

implicit function A function that is implied by an equation, rather than given explicitly as y = f(x). For example, the equation 2x + 3y = 4 gives y as an implicit function of x. By solving this equation for y, we get y as an explicit function of x, namely y = (−2/3)x + 4/3.

As a second example, the equation $x^2 + y^2 = 1$ (the equation of a unit circle) defines *two* explicit functions of x, namely $y = \sqrt{1 - x^2}$ and $y = -\sqrt{1 - x^2}$ (the upper and lower half of the unit circle, respectively). In other cases it may be difficult or impossible to find y explicitly as a function of x, as in the example $x^3 + y^3 = 3xy$. Still, such an equation does define y as one or several functions of x.

See also IMPLICIT DIFFERENTIATION.

improper integral A definite integral $\int_a^b f(x)\,dx$ whose interval of integration (a, b) is unbounded (that is, either a or b or both are at infinity), or whose integrand $f(x)$ has a finite number of infinite discontinuities on (a, b). Such an integral can only be evaluated as a limit. For example, $\int_0^\infty e^{-x}\,dx = \lim_{b\to\infty} [\int_0^b e^{-x}\,dx] = \lim_{b\to\infty} [1 - e^{-b}] = 1$, because $\lim_{b\to\infty} e^{-b} = 0$ for $b > 0$. As another example, $\int_0^1 1/x^2\,dx$ is an improper integral, because $1/x^2$ becomes infinite at $x = 0$, which is in the interval of integration. In this case we attempt to compute $\lim_{a\to0} \int_a^1 1/x^2\,dx = \lim_{a\to0} (1/a - 1)$ but since $1/a$ becomes infinite as $a \to 0$, the limit does not exist; we say that the integral *diverges*.

increasing function *See* FUNCTION, INCREASING.

increment A change in the value of a variable. If the variable is x, we denote its increment by Δx, or by $x_2 - x_1$.

indefinite integral *See* INTEGRAL, INDEFINITE.

independent variable The variable x in the expression $y = f(x)$. We can change the value of x at will (as long as it is taken from the domain of f); by contrast, the value of y will be determined by our choice of x and is therefore called the *dependent variable.*

indeterminate form An algebraic expression that takes the form $0/0$ or ∞/∞ when the independent variable assumes certain values. For example, the expression $(1 - \cos x)/x^2$ takes the form $0/0$ when $x = 0$. This usually indicates that the value of the expression must be obtained indirectly through an algebraic simplification or through a limiting process. In the example given, we can multiply and divide by $(1 + \cos x)$, giving us $(1 - \cos x)(1 + \cos x)/[(1 + \cos x)x^2] = (1 - \cos^2 x)/[(1 + \cos x)x^2]$ $= \sin^2 x/[(1 + \cos x)x^2] = [(\sin x)/x]^2/(1 + \cos x)$. As $x \to 0$, the numerator has the limit 1 and the denominator has the limit 2, so the entire expression has the limit $1/2$. Other indeterminate forms are $\infty \cdot 0$, 1^∞, 0^0, ∞^0, and $\infty - \infty$, but these can be transformed into the forms $0/0$ or ∞/∞ by various algebraic processes. For example, the expression $y = x^x$, defined for $x > 0$, takes the form 0^0 when $x = 0$, so we take the natural logarithm of both sides and get $\ln y = x \ln x = (\ln x)/(1/x)$, which takes the form $(-\infty)/\infty$ at $x = 0$ (it can be shown that $\lim_{x\to0} x^x = 1$ as $x \to 0$ through positive values).
See also L'HOSPITAL'S RULE.

index of summation The subscript i in the expression $\sum_{i=1}^{n} a_i$. The choice of letter for the summation index is immaterial; that is, $\sum_{i=1}^{n} a_i = \sum_{j=1}^{n} a_j$; for this reason, the summation index is sometimes called a *dummy index*. *See also* DUMMY VARIABLE.

inequality, triangle *See* TRIANGLE INEQUALITY.

infinite discontinuity *See* DISCONTINUITY, INFINITE.

infinite interval *See* INTERVAL, INFINITE.

infinite limit A function f whose values become arbitrarily large as the independent variable gets closer and closer to a number c. More precisely, we say that $\lim_{x \to c} f(x) = \infty$ or $-\infty$ if for every positive number M, no matter how large, there exists a number δ (Greek delta) such that $|f(x)| > M$ whenever $|x - c| < \delta$; that is, we can make the absolute value of f(x) as large as we please by getting sufficiently close to c from either side. For example, $\lim_{x \to 1} 1/(1 - x^2) = \infty$, because we can make the value of $|1/(1 - x^2)|$ as large as we please by letting x be sufficiently close to 1.

infinite series

Convergence of, *See* CONVERGENCE OF A SERIES.
Divergence of, *See* DIVERGENCE OF A SERIES.
Geometric, *See* GEOMETRIC SERIES.
Harmonic, *See* HARMONIC SERIES.
Partial sums of: The sequence of sums a_1, $(a_1 + a_2)$, $(a_1 + a_2 + a_3)$, ... of the terms of an infinite sequence $a_1, a_2, a_3, \ldots$.
Telescopic, *See* TELESCOPIC SERIES.

infinity, limit at The limit of a function f as the independent variable tends to infinity or negative infinity; that is, $\lim_{x \to \infty} f(x)$ or $\lim_{x \to -\infty} f(x)$ (provided either of these limits exists). For example, the limit of $f(x) = (2x - 1)/(3x + 2)$ as $x \to \infty$ is 2/3, so we write $\lim_{x \to \infty} (2x - 1)/(3x + 2) = 2/3$ (this is also the limit as $x \to -\infty$).
See also ASYMPTOTE, HORIZONTAL.

inflection point A point where the graph of a function changes its concavity from concave up to concave down or vice versa. If the graph has the equation $y = f(x)$, then an inflection point occurs at a point $x = c$ where $f''(c)$ (the second derivative of f at $x = c$) is either zero or undefined. For example, the functions $y = x^3$ and $y = x^{1/3}$ both have an inflection point at $x = 0$; for the first function, $y'' = 0$ at $x = 0$, while for the second function, y'' is undefined there.

Inflection point: (a) y″ = 0
at 0; (b) y″ undefined at 0

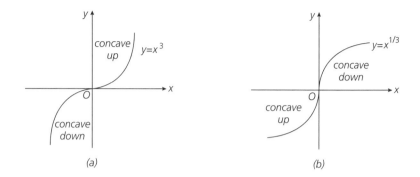

(a)   (b)

initial condition The value of a function at a given point of its domain. This
information is needed when solving a differential equation in order to
get a particular solution.
See also DIFFERENTIAL EQUATION.

instantaneous rate of change *See* DERIVATIVE; RATE OF CHANGE,
INSTANTANEOUS.

integral Definite: the limit of a Riemann sum as the number of subdivisions
tends to infinity and the length of each subinterval tends to zero. If
the function inside the integral sign is f(x) and the interval of
integration is [a, b], we denote the definite integral by $\int_a^b f(x)dx$ (read:
"the definite integral of f(x) from *a* to *b*"). The definite integral is a
number. If $f(x) \geq 0$ on the entire interval [a, b], we can interpret this
number as the area under the graph of f(x) from x = a to x = b.
See also RIEMANN SUM.

Indefinite: an antiderivative of f(x); that is, a function F(x) whose
derivative is f(x). We denote an indefinite integral of f(x) by $\int f(x)\, dx$.
Because the derivative of a constant is zero, it follows that F(x) + C is
also an antiderivative of f(x), so we write $\int f(x)\, dx = F(x) + C$; *C* is
called the *constant of integration*. The phrase "indefinite integral"
indicates that an antiderivative of f(x) can be determined only up to an
arbitrary constant of integration. For example, $\int x^2\, dx = x^3/3 + C$,
because $d/dx(x^3/3 + C) = x^2$.
See also ANTIDERIVATIVE.

integrand The function under an integral sign. For example, the integrand of
$\int x^2\, dx$ is x^2 (the dx is *not* considered part of the integrand).

integration The process of finding an integral (definite or indefinite).
By parts:

For definite integrals: $\int_a^b f(x)g'(x)\,dx = f(x)g(x)\big|_a^b - \int_a^b f'(x)g(x)\,dx$

For indefinite integrals: $\int f(x)g'(x)\,dx = f(x)g(x) - \int f'(x)g(x)\,dx$

See also INTEGRATION BY PARTS for examples of this rule.

Limits of: The endpoints x = a and x = b of the interval of integration [a, b]; x = a is the *lower limit,* and x = b the *upper limit* (these should not be confused with the limit of a Riemann sum as the number of subintervals tends to infinity). We use the symbol $\int_a^b$ to indicate the limits of integration.

Numerical: *See* NUMERICAL INTEGRATION.

Rules of:

Definite integration: $\int_a^b f(x)\,dx = F(b) - F(a)$, where $F'(x) = f(x)$

$$d/dx \int_a^x f(t)\,dt = f(x)$$

$$\int_a^b c\,f(x)\,dx = c \int_a^b f(x)\,dx, \text{ where } c = \text{constant}$$

$$\int_a^b [f(x) \pm g(x)]\,dx = \int_a^b f(x)\,dx \pm \int_a^b g(x)\,dx$$

$$\int_a^b f(x)\,dx + \int_b^c f(x)\,dx = \int_a^c f(x)\,dx$$

$$\int_a^b f(x)\,dx = - \int_b^a f(x)\,dx$$

$$\int_a^a f(x)\,dx = 0$$

If $f(x) \leq g(x)$ on [a, b], then

$$\int_a^b f(x)\,dx \leq \int_a^b g(x)\,dx$$

If $m \leq f(x) \leq M$ on [a, b], then

$$m(b - a) \leq \int_a^b f(x)\,dx \leq M(b - a)$$

Indefinite integration: $\int f(x)\,dx = F(x) + C$, where $F'(x) = f(x)$ and C = an arbitrary constant

$$d/dx \int f(x)\,dx = f(x)$$

$$\int c\,f(x)\,dx = c \int f(x)\,dx, \text{ where } c = \text{constant}$$

$$\int [f(x) \pm g(x)]\,dx = \int f(x)\,dx \pm \int g(x)\,dx$$

By substitution:

For definite integrals:

$$\int_a^b f(g(x))g'(x)\,dx = F(g(x))\big|_a^b, \text{ where } F'(g(x)) = f(g(x))$$

Alternate form:

$$\int_a^b f(u)\,du = F(u)\big|_{g(a)}^{g(b)}, \text{ where } F'(u) = f(u) \text{ and } u = g(x)$$

For indefinite integrals:

$$\int f(g(x))\, g'(x)\, dx = F(g(x)) + C, \text{ where } F'(g(x)) = f(g(x))$$
and C = an arbitrary constant.

Alternate form:

$$\int f(u)\, du = F(u) + C, \text{ where } F'(u) = f(u) \text{ and } u = g(x);$$
C = an arbitrary constant.

See also SUBSTITUTION, METHOD OF for examples of integration by substitution.

integration formulas *See* Section Four C.

integration by parts A method of finding integrals (definite or indefinite), when the integrand (the function inside the integral) is of the form $f(x)g'(x)$, that is, the product of one function and the derivative of another function.

For definite integrals: $\int_a^b f(x)g'(x)\, dx = f(x)g(x)\Big|_a^b - \int_a^b f'(x)g(x)\, dx$

For indefinite integrals: $\int f(x)g'(x)\, dx = f(x)g(x) - \int f'(x)g(x)\, dx$

As an example, let us find $\int x \sin x\, dx$. We write $f(x) = x$, $g'(x) = \sin x$, so that $f'(x) = 1$, $g(x) = -\cos x$ (we will combine the constant of integration with that of the second integral). Thus $\int x \sin x\, dx = -x \cos x - \int 1 \cdot (-\cos x)\, dx = -x \cos x + \int \cos x\, dx = -x \cos x + \sin x + C$.
The decision which function we should call $f(x)$ and which $g'(x)$ is dictated by the ease of finding $g(x)$ (the antiderivative of $g'(x)$), the goal being to have a new integral which is easier to find than the original. In the example just given, had we chosen to write $f(x) = \sin x$, $g'(x) = x$, we would have ended up with $(x^2/2)\sin x - \int (x^2/2) \cos x\, dx$, making the new integral more difficult to find than the original.

integration by substitution *See* SUBSTITUTION, METHOD OF.

intercept A point at which the graph of a function $y = f(x)$ crosses the *x*- or *y*-axis.
X-intercept: the point where the graph crosses the *x*-axis; its coordinates are $(x, 0)$, where x is a solution of the equation $f(x) = 0$.
Y-intercept: the point where the graph crosses the *y*-axis; its coordinates are $(0, f(0))$.

interest The amount of money paid for getting a loan. The "loan" may be an investment made when one opens a bank account, in which case the investor gives a loan to the bank.

Compound: *See* COMPOUND INTEREST.
Continuous: *See* COMPOUND INTEREST, CONTINUOUS.
Simple: *See* SIMPLE INTEREST.

Intermediate Value Theorem Let f be a continuous function on the closed interval [a, b], and let m be a number between f(a) and f(b) (that is, f(a) ≤ m ≤ f(b)); then there exists at least one number c on [a, b] for which f(c) = m. Note: The requirement that f is continuous on [a, b] is crucial. Also, there may be more than one value c for which f(c) = m.

interval Generally, a segment of the number line.
   Closed: An interval that includes its two endpoints. We use the notation [a, b].
   Half-open: An interval that includes one endpoint but not the other. We use the notation [a, b) or (a, b], depending on which endpoint is included.
   Infinite: An interval in which one or both "endpoints" are at infinity (note that we put the word "endpoints" between quotation marks, because infinity is not a point on the number line and therefore cannot be a true endpoint of an interval). Five kinds of

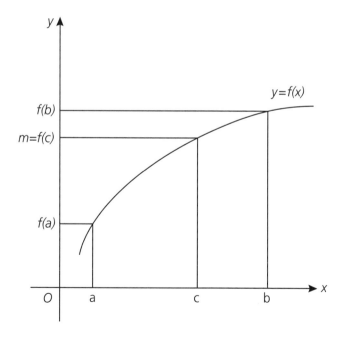

Intermediate Value
Theorem

Interval, showing two notations: (a) closed: [a, b]; (b) half-open: [a, b) and (a, b]; (c) infinite: (−∞, a) and [a, ∞); (d) open: (a, b)

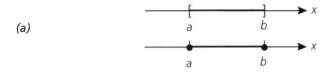

(a)

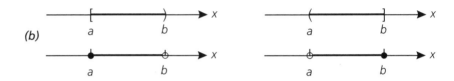

(b)

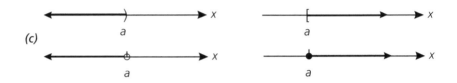

(c)

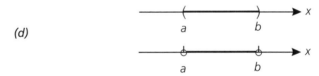

(d)

infinite intervals are possible: (−∞, a], (−∞, a), [a, ∞), (a, ∞), and (−∞, ∞) (the last is the entire number line). Note that the symbol ∞ is always enclosed by an open parenthesis.

Open: An interval that includes neither of its two endpoints. We use the notation (a, b). (Note: this should not be confused with the notation (a, b) for coordinates; the correct meaning will usually be clear from the context.)

interval of convergence The interval on which an infinite series converges. For example, the geometric series $1 + x + x^2 + \ldots$ converges for all values of x in the open interval (−1, 1) (but not at its endpoints).
  See also POWER SERIES.

interval of integration The interval [a, b] over which a definite integral $\int_a^b f(x)\,dx$ is evaluated.
  See also INTEGRAL, DEFINITE.

inverse function Let f be a one-to-one function. Its inverse, denoted by f^{-1}, is the function that satisfies the equations $f^{-1}(f(x)) = x$ for all values of x in the domain of f, and $f(f^{-1}(x)) = x$ for all values of x in the domain of f^{-1}. For example, the function $y = f(x) = x^2$, $x \geq 0$ has an inverse $y = f^{-1}(x) = \sqrt{x}$. Note that the domain of f is the range of f^{-1}, and vice versa. Not every function has an inverse; for example, the function $y = f(x) = x^2$ (defined for all real x) has no inverse, because it fails the Horizontal Line Test. A function has an inverse if and only if it is *one-to-one* and *onto*.

See also ONE-TO-ONE FUNCTION; ONTO FUNCTION; HORIZONTAL LINE TEST.

inverse function, derivative of Let $y = f(x)$ be a one-to-one function. Its inverse is $x = f^{-1}(y) = g(y)$. The derivative of the inverse function (with y the independent variable and x the dependent variable) is $g'(y) = dx/dy$. Thinking of dx and dy as differentials (that is, as ordinary algebraic quantities), we can write this as $1/(dy/dx)$, or $1/f'(x)$. We thus have the nice rule $g'(y) = 1/f'(x)$, where f and g are inverses of each other.

As an example, consider the function $y = \sin x$ in the interval $[-\pi/2, \pi/2]$. Its inverse is $x = \arcsin y$. We thus have $d(\arcsin y)/dy = 1/[d(\sin x)/dx] = 1/\cos x = 1/\sqrt{1 - \sin^2 x} = 1/\sqrt{1 - y^2}$; we take the positive square root because $\cos x$ is positive on the interval $(-\pi/2, \pi/2)$; of course, at the endpoints $-\pi/2$ and $\pi/2$, $\cos x$ is zero so the derivative is undefined. If we wish, we can change the variable in the last formula to x and write the formula as $d(\arcsin x)/dx = 1/\sqrt{1 - x^2}$. The formula illustrates the usefulness of the LEIBNIZ NOTATION for the derivative of a function.

irrational number A number that cannot be written as the ratio of two integers. Examples are $\sqrt{2}$, π, and e (the base of natural logarithms). The decimal expansion of an irrational number is nonterminating and nonrepeating; that is, its digits never end and never repeat in the same order.

See also RATIONAL NUMBER.

iteration A computational procedure in which a number is found from a previous number by repeatedly using the same formula. For example, the *Fibonacci numbers* 1, 1, 2, 3, 5, 8, 13, 21, . . . (each number from the third on is the sum of its two predecessors) can be generated by the iteration formula $a_1 = a_2 = 1$, $a_n = a_{n-2} + a_{n-1}$, $n = 3, 4, 5, \ldots$.

latus rectum *See* ELLIPSE; PARABOLA.

Law of Cosines A theorem from trigonometry that says: In any triangle,

$$a^2 = b^2 + c^2 - 2bc \cos A$$

where A is the angle included between the sides b and c. Similarly,

$$b^2 = c^2 + a^2 - 2ca \cos B$$
$$c^2 = a^2 + b^2 - 2ab \cos C$$

Law of Sines A theorem from trigonometry that says: In any triangle,

$$a/\sin A = b/\sin B = c/\sin C = 2R$$

where A, B, and C are the angles opposite sides a, b, and c, respectively, and R is the radius of the circle inscribing the triangle.

left-handed limit *See* LIMIT, ONE-SIDED.

Leibniz notation The notation dy/dx for the derivative of a function $y = f(x)$ with respect to x. Also known as the "d-notation" and named after Gottfried Wilhelm Leibniz, coinventor with Newton of the calculus (*see* the biographical section). It is the limit of the difference quotient $\Delta y/\Delta x$ as Δx and Δy tend to zero. Although not a ratio itself, the symbol dy/dx behaves as if it were an ordinary algebraic fraction. For example, the *Chain Rule* says that if $y = f(u)$ and $u = g(x)$, then the derivative of the composite function $y = f(g(x)) = h(x)$ with respect to x is obtained by multiplying the derivatives of the two component functions: $dy/dx = (dy/du)(du/dx) = f'(u)g'(x)$. It is this ease of operation that has made Leibniz's notation so convenient in calculus.
See also DERIVATIVE; DIFFERENCE QUOTIENT.

Leibniz's rule A rule for finding the higher derivatives of a product of two functions f and g:

$$(fg)'' = fg'' + 2f'g' + f''g$$
$$(fg)''' = fg''' + 3f'g'' + 3f''g' + f'''g$$
$$(fg)^{(4)} = fg^{(4)} + 4f'g''' + 6f''g'' + 4f'''g' + f^{(4)}g$$

and so on. The *n*th derivative of *fg* follows the same rule as the binomial expansion of $(a + b)^n$, with derivatives replacing the exponents.
See also BINOMIAL THEOREM.

lemniscate A closed looping curve resembling the infinity symbol ∞. Its rectangular equation is $(x^2 + y^2) = a^2(x^2 - y^2)$; its polar equation is $r^2 = a^2 \cos 2\theta$.

length Of an arc, *See* ARC LENGTH.
Of an interval [a, b]: $b - a$.

L'Hospital's Rule A rule that allows us to find the limit of indeterminate expressions of the form 0/0 or ∞/∞ (or expressions that can be reduced to these forms). The rule says: Let f and g be two differentiable functions on an open interval (a, b), except possibly at a point c in (a, b). If $\lim_{x \to c} f(x)/g(x)$ leads to the form 0/0 or ∞/∞, then $\lim_{x \to c} f(x)/g(x) = \lim_{x \to c} f'(x)/g'(x)$ (note that the last expression involves the *ratio of the derivatives;* this should not be confused with the Quotient Rule, which involves the derivative of a ratio).

For example, if we attempt to find $\lim_{x \to 0} (e^x - 1)/x$ by directly substituting x = 0, we get 0/0, but using L'Hospital's Rule with $f(x) = e^x - 1$ and $g(x) = x$, we get $\lim_{x \to 0} e^x/1 = 1$. The rule is named after GUILLAUME L'HOSPITAL (sometimes spelled L'Hopital).

See also INDETERMINATE FORM.

limit At infinity: If the values of f(x) approach a number L as x gets larger and larger, we write $f(x) \to L$ as $x \to \infty$, or briefly $\lim_{x \to \infty} f(x) = L$. For example the function $y = f(x) = e^{-x}$ has the limit 0 as $x \to \infty$, so we write $\lim_{x \to \infty} e^{-x} = 0$. A similar definition applies when $x \to -\infty$ (for example, $\lim_{x \to -\infty} e^x = 0$). In either case, the graph of f(x) has a *horizontal asymptote* y = L.

See also ASYMPTOTE, HORIZONTAL.

Infinite: If the values of f(x) get larger and larger as x approaches a number c, we say that f(x) approaches infinity and write $f(x) \to \infty$ as $x \to c$, or briefly $\lim_{x \to c} f(x) = \infty$. For example, the function $y = f(x) = 1/x^2$ approaches infinity as $x \to 0$, so we write $\lim_{x \to 0} 1/x^2 = \infty$. If the values of f(x) are negative and get larger and larger in absolute value as x approaches c, we write $f(x) \to -\infty$ as $x \to c$, or briefly $\lim_{x \to c} f(x) = -\infty$. (The two cases can be combined by saying that f(x) becomes infinite as $x \to c$ if |f(x)| gets larger and larger as we get close to c.) In either case, the graph of f(x) has a *vertical asymptote* at x = c.

See also ASYMPTOTE, VERTICAL.

Of a function: Intuitively, a function f(x) approaches a limit L as x approaches a number c if we can make the values of f(x) get arbitrarily close to L by letting x get sufficiently close to c. More formally, f(x) approaches a limit L as x approaches c if, for every positive number ε (Greek epsilon), no matter how small, there exists a number δ (Greek delta) such that $|f(x) - L| < ε$ whenever $|x - c| < δ$. We write $f(x) → L$ as $x → c$, or $\lim_{x→c} f(x) = L$. Note that the function need not be defined *at* $x = c$, only *near c*. For example, the function $f(x) = (x^2 - 1)/(x - 1)$ is defined for all x except $x = 1$, but its limit as $x → 1$ is 2.

Of a sequence: We say that the sequence $a_1, a_2, a_3, \ldots a_i, \ldots$ has a limit L if, for every positive number ε (Greek epsilon), no matter how small, there exists an integer N such that $|a_i - L| < ε$ whenever $i > N$. This means that we can make the members of the sequence get as close to L as we wish, by going out far enough along the sequence (that is, by choosing a sufficiently large i). We write $a_i → L$ as $i → ∞$, or $\lim_{i→∞} a_i = L$. For example, the sequence $2/1, 3/2, 4/3, \ldots, (i + 1)/i$, $\ldots$ has the limit 1 as $i → ∞$, so we write $\lim_{i→∞}(i + 1)/i = 1$.

Of a series: We say that the infinite series $a_1 + a_2 + a_3 + \ldots$ has a limit S if the sequence of *partial sums* $a_1, (a_1 + a_2), (a_1 + a_2 + a_3), \ldots, (a_1 + a_2 + a_3 + \ldots + a_i), \ldots$ approaches S as the number of terms i grows to infinity. We write $a_1 + a_2 + a_3 + \ldots + a_i → S$ as $i → ∞$, or simply $a_1 + a_2 + a_3 + \ldots + a_i + \ldots = S$; this is often abbreviated by using the sigma notation: $\sum_{i=1}^{∞} a_i = S$. For example, the geometric series $1 + 1/2 + 1/4 + 1/8 + \ldots$ approaches the limit 2 as the number of terms grows to infinity, so we write $\sum_{i=0}^{∞} 1/2^i = 2$ (note that here i starts at 0, because the first term is $1 = 1/2^0$).

One-sided: A function $f(x)$ approaches a *one-sided* limit L as x approaches a number c if we can make the values of $f(x)$ get arbitrarily close to L by letting x get sufficiently close to c *from one side of c*. Depending on whether the approach is from the left or right of c, we call the limit a *left-handed* or *right-handed* limit, and write $f(x) → L$ as $x → c^-$ or $x → c^+$, respectively. One-sided limits occur when a function is defined on an open interval (a, b), but not at one or both of its endpoints. For example, the function $f(x) = \ln x$ (the natural logarithmic function) is defined on the interval $(0, ∞)$, and $\ln x → -∞$ as $x → 0^+$.

limits of integration The endpoints of the interval [a, b] over which we compute a definite integral of a function f(x). We write $\int_a^b f(x)\, dx$; the numbers *a* and *b* are the *lower* and *upper limits of integration,* respectively.

line(s), equations of General form: The equation Ax + By = C, where *A, B,* and *C* are constants, with *A* and *B* not both zero. Also known as the *standard linear equation,* or *general linear equation in two variables.*

Horizontal: The equation y = a, where *a* is a constant.

Parallel: Two non-vertical lines are parallel if and only if they have the same slope; their equations are y = mx + b and y = mx + c. If the lines are vertical, their slope is undefined.

Perpendicular: Two lines are perpendicular if their slopes are *negative reciprocals* of each other; that is, if one slope is *m*, then the other is $-1/m$. If one of the lines is horizontal, the other will be vertical and its slope is undefined; in this case the equations of the lines are y = a and x = b, respectively.

Point-slope form: The equation $y - y_1 = m(x - x_1)$, where *m* is the slope and (x_1, x_2) a point on the line. For example, the point-slope equation of the line with slope 2 and passing through the point $(1, -3)$ is $y - (-3) = 2(x - 1)$, that is, $y + 3 = 2(x - 1)$; this is usually simplified and written in the slope-intercept form, $y = 2x - 5$.

Slope-intercept form: The equation y = mx + b, where *m* is the slope and (0, b) the *Y*-intercept of the line.

Vertical: The equation x = b, where *b* is a constant.

linear approximation An approximation in which the value of a function y = f(x) near a point x = c is replaced by the *y*-value (that is, the height) of the tangent line to the graph of *f* at *c*. Using the point-slope equation of a line and the fact that the tangent line at *c* has slope $m = f'(c)$, the linear approximation formula is $y - f(c) \approx f'(c)(x - c)$, or $f(x) \approx f(c) + f'(c)(x - c)$. For example, the linear approximation to $f(x) = \sqrt{x}$ near x = 1 is $\sqrt{x} \approx \sqrt{1} + (1/2\sqrt{1})(x - 1) = 1 + (x - 1)/2 = (x + 1)/2$. We can use this formula to approximate square roots of numbers near x = 1; for example, $\sqrt{1.1} \approx (1.1 + 1)/2 = 1.05$, compared to the true value 1.0488 . . . . The approximation gets better the closer *x* is to 1, or generally, to *c*.

linear combination An expression of the form $a_1x_1 + a_2x_2 + \ldots + a_nx_n$, where the a_i are constants (coefficients) and the x_i are either variables or functions of a variable. For example, A cos x + B sin x is a linear combination of the functions cos x and sin x.

linear differential equation A differential equation in which only first powers of the unknown function *y* and its derivatives show up. For example, the equation $x^2y'' + 2xy' + y = \sin x$ is linear. Examples of nonlinear equations are $xy' + y^2 = \ln x$ and $xyy' + 2y = 0$ (the latter because of the appearance of the product yy'). Linear differential equations are, generally speaking, among the simplest to solve.

linear function A function of the form f(x) = ax + b, where *a* and *b* are constants. Because its graph is a straight line, the equation is often written in the form y = mx + b, where *m* is the slope of the line and (0, b) is its *Y*-intercept.

linear spiral *See* ARCHIMEDES, SPIRAL OF.

logarithm The exponent to which a fixed positive number *b* (excluding 1), called the *base,* must be raised to give another positive number; that is, $b^x = c$ if and only if $x = \log_b c$, read "logarithm to the base *b* of *c*." For example, each of the following exponential equations can be written as a logarithm: $10^2 = 100 \Rightarrow \log_{10} 100 = 2$; $10^3 = 1{,}000 \Rightarrow \log_{10} 1{,}000 = 3$; $10^1 = 10 \Rightarrow \log_{10} 10 = 1$, $10^0 = 1 \Rightarrow \log_{10} 1 = 0$; and $10^{-1} = 1/10 = 0.1 \Rightarrow \log_{10} 0.1 = -1$. If the number *c* is not an exact power of *b,* then the logarithm will not be an integer, and it can only be found approximately. For example, $10^{0.69897} = 5$ (approximately), so $0.69897 = \log_{10} 5$ (approximately). In principle, any positive number other than 1 can be used as a base; for example, $\log_2 5 \approx 2.32193$ because $2^{2.32193} \approx 5$. In practice, however, two standard bases are being used: 10, and the irrational number *e,* whose approximate value is 2.7182818. *See* e.

Common: Logarithms to base 10, denoted by log. For example, $\log 5 \approx 0.69897$, because $10^{0.69897} \approx 5$.

Natural: Logarithms to base *e,* denoted by ln. For example, $\ln 5 \approx 1.60944$, because $e^{1.60944} \approx 5$.

See also CHANGE OF BASE.

logarithmic function The function $y = f(x) = \log_b x$, where the base *b* is a positive number different from 1. By definition, $y = \log_b x$ if and only if $x = b^y$; hence the logarithmic function is the inverse of the exponential function. Its domain is $(0,\infty)$, that is, all positive real numbers, and its range is $(-\infty,\infty)$ Its graph has the y-axis as a vertical asymptote.

Natural: The logarithmic function with base e ≈ 2.7182818. It is denoted by ln. That is, $y = \ln x$ if and only if $x = e^y$.

Properties of: Let $f(x) = \log_b x$. Then $f(1) = 0$, $f(x_1 x_2) = f(x_1) + f(x_2)$, $f(x_1/x_2) = f(x_1) - f(x_2)$, $\lim_{x \to 0^+} f(x) = -\infty$ when b > 1 and ∞ when $0 < b < 1$, and $f'(x) = 1/(\ln b)x$. If in particular b = e, we have $f'(x) = 1/x$; that is, d/dx ln x = 1/x. Furthermore, if x < 0, then $-x > 0$, so the function ln (−x) is defined. Using the Chain Rule, its derivative is d/dx ln (−x) = $[1/(-x)] \cdot (-1) = 1/x$. Thus the functions ln x for x > 0 and ln (−x) for x < 0 have the same derivative, 1/x. It is customary to combine these statements into a single formula, d/dx ln |x| = 1/x.

See also EXPONENTIAL FUNCTION.

logarithmic spiral A spiral curve whose polar equation is $r = e^{a\theta}$, where *e* is the base of natural logarithm. This curve often appears in nature, as

Logarithmic function

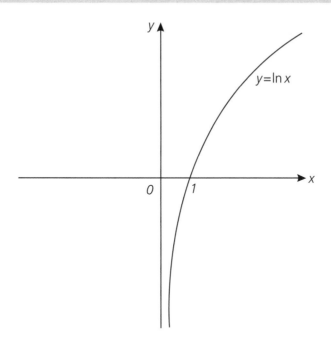

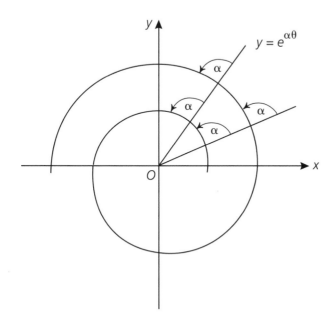

Logarithmic spiral

for example in the nautilus shell or the seed arrangement of a sunflower. It has the property that any straight line through the center intercepts the spiral at the same constant angle $\alpha = \cot^{-1} a$.

Maclaurin series If a function f(x) has derivatives of all orders at x = 0, it can be represented by the infinite series $f(0) + f'(0)x + f''(0)x^2/2! + \ldots + f^{(n)}(0)x^n/n! + \ldots$, where $f^{(n)}$ is the *n*th derivative of *f,* and n! (read "*n* factorial") is the product $1 \cdot 2 \cdot 3 \cdot \ldots \cdot n$.

This series is called the *Maclaurin series* of f(x), named after COLIN MACLAURIN; we write $f(x) = \sum_{i=0}^{\infty} f^{(i)}(0)x^i/i!$.

See also TAYLOR POLYNOMIAL; TAYLOR SERIES; TAYLOR THEOREM.

major axis *See* ELLIPSE.

marginal cost A concept in mathematical economics. If C(x) denotes the cost of producing *x* units of a commodity, then the *marginal cost* is the expression $MC(x) = C(x + 1) - C(x)$. [Note: *MC* is one symbol, not a product.] This expresses the cost of producing *one additional unit,* when *x* units are already being produced. Because *x* is a discrete variable (that is, it assumes only integer values), the marginal cost function is discontinuous; however, for large *x* values, we can approximate it by the derivative of the cost function: $MC(x) = [C(x + 1) - C(x)]/1 \approx dC/dx = C'(x)$.

marginal profit A concept in mathematical economics. If P(x) denotes the profit when producing and selling *x* units of a commodity, then the *marginal profit* is the expression $MP(x) = P(x + 1) - P(x)$. [Note: *MP* is one symbol, not a product.] This expresses the profit from producing and selling *one additional unit,* when *x* units are already being produced and sold. Because *x* is a discrete variable (that is, it assumes only integer values), the marginal profit function is discontinuous; however, for large *x* values, we can approximate it by the derivative of the profit function: $MP(x) = [P(x + 1) - P(x)]/1 \approx dP/dx = P'(x)$.

marginal revenue A concept in mathematical economics. If R(x) denotes the revenue when producing and selling *x* units of a commodity, then the *marginal revenue* is the expression $MR(x) = R(x + 1) - R(x)$. [Note: *MR* is one symbol, not a product.] This expresses the revenue of producing and selling *one additional unit,* when *x* units are already being produced and sold. Because *x* is a discrete variable (that is, it assumes only integer values), the marginal revenue function is discontinuous; however, for large *x* values, we can approximate it by the derivative of the revenue function: $MR(x) = [R(x + 1) - R(x)]/1 \approx dR/dx = R'(x)$.

maximum Absolute (also called *global*): The largest value of a function *f* in its entire domain. More formally, f(x) has an absolute maximum at x = c if f(c) ≥ f(x) for all *x* in the domain of *f*.

 Relative (also called *local*): The largest value of a function *f* on an open interval (a, b). More formally, f(x) has a relative maximum at x = c (where a < c < b) if f(c) ≥ f(x) for all *x* in (a, b).

mean, arithmetic *See* ARITHMETIC MEAN.

mean, geometric *See* GEOMETRIC MEAN.

Mean Value Theorem Let *f* be a continuous function on a closed interval [a, b] and differentiable (that is, has a derivative) on the open interval (a, b). Then there exists at least one number *c* in (a, b) for which f′(c) = [f(b) – f(a)]/(b – a). Geometrically, this means there is at least one point between *a* and *b* where the tangent line to graph of *f* has the same slope as the *secant line* connecting the points (a, f(a)) and (b, f(b)). (Note: there may be more than one such point). This is an example of an *existence theorem:* it guarantees that a point with the prescribed property exists, but it does not tell us how to find it.

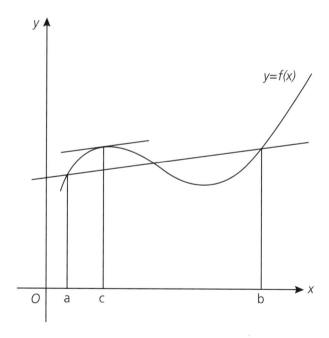

Mean Value Theorem

midpoint The midpoint of the line segment connecting the points (x_1, y_1) and (x_2, y_2) is the point $((x_1 + x_2)/2, (y_1 + y_2)/2)$. For example, the midpoint of the line segment connecting the points $(2, 5)$ and $(3, -8)$ is $(5/2, -3/2)$.

Midpoint Rule A procedure for approximating the value of a definite integral:
$$\int_a^b f(x)\, dx \approx \sum_{i=1}^{n} [f(x_{i-1} + x_i)/2]\Delta x,$$ where x_i, $i = 1, 2, \ldots, n$, are the points of subdivision of the interval $[a, b]$ into n equal subintervals, each of length $\Delta x = (b - a)/n$.

See also DEFINITE INTEGRAL; RIEMANN SUM; SIMPSON'S RULE; TRAPEZOID RULE.

minimum Absolute (global): The smallest value of a function f in its entire domain. More formally, $f(x)$ has an absolute minimum at $x = c$ if $f(c) \leq f(x)$ for all x in the domain of f.

Relative (local): The smallest value of a function f on an open interval (a, b). More formally, $f(x)$ has a relative minimum at $x = c$ (where $a < c < b$) if $f(c) \leq f(x)$ for all x in (a, b).

minor axis See ELLIPSE.

model, mathematical A term referring to any "real-life" situation that can be described by a set of equations (such as differential equations), or by some other mathematical expression (for example, a set of inequalities). The process is called *mathematical modeling*.

modeling The processes of translating a real-life problem into a set of equations or inequalities whose solution gives an approximate description of the original problem. This usually involves making various assumptions on the original problem, intended to simplify the mathematical treatment. As an example, the to-and-fro oscillations of a swing, assuming the absence of friction, can be described ("modeled") by the differential equation $y'' + k^2y = 0$, where y (the angular deviation from the equilibrium position) is a function of the time t and k is a constant. Its solution, $y = A \cos kt + B \sin kt$ is an approximate mathematical description of the ensuing motion. By adding a term ry' to the equation (where r is another constant), we can take into account the resistance to the motion due to the presence of air, thereby making the model more accurate.

monotonic Function: A function that is either increasing or decreasing on an entire open interval. For example, the function $y = x^2$ is monotone-increasing on any interval $(0, a)$ where $a > 0$, and monotone-decreasing on any interval $(b, 0)$ where $b < 0$.

Sequence: A sequence $a_1, a_2, a_3, \ldots$ for which either $a_1 < a_2 < a_3 < \ldots$ (monotone-increasing sequence), or $a_1 > a_2 > a_3 > \ldots$ (monotone-decreasing sequence). For example, the sequence $2/1, 3/2, 4/3, \ldots$ is monotone-decreasing.

multiplication of two functions The product of two functions f and g, written fg. That is to say, $(fg)(x) = f(x)\,g(x)$. For example, if $f(x) = 2x + 1$ and $g(x) = 3x - 2$, then $(fg)(x) = (2x + 1)(3x - 2) = 6x^2 - x - 2$.

natural exponential function *See* EXPONENTIAL FUNCTION, NATURAL.

natural logarithm *See* LOGARITHM, NATURAL.

natural logarithmic function *See* LOGARITHMIC FUNCTION, NATURAL.

neighborhood of a point An open interval containing the point.

Newton's method (also known as Newton-Raphson method) An iterative procedure for finding the approximate zeros of a function f, that is, the solutions of the equation $f(x) = 0$. Let c be a zero of f, and assume that f is differentiable in a neighborhood of c. We perform the following steps:

Step 1: Make an initial guess for the value of c; call it x_0.
Step 2: Find a better approximation x_1 for c from the formula $x_1 = x_0 - f(x_0)/f'(x_0)$.
Step 3: Repeat step 2 with x_1 replacing x_0, obtaining a second approximation x_2.
Step 4: Continue in this manner, using the iterative formula $x_{n+1} = x_n - f(x_n)/f'(x_n)$, until the desired accuracy is obtained.

As an example, suppose we want to approximate the value of $\sqrt{2}$. We let $f(x) = x^2 - 2$ and seek to find the positive zero of this function, that is, the positive solution of the equation $f(x) = 0$. We have $f(x_n) = x_n^2 - 2$ and $f'(x_n) = 2x_n$. Putting this in the iterative formula and simplifying, we get $x_{n+1} = (1/2)(x_n + 2/x_n)$. We now make an initial guess, say $x_0 = 1.5$, and put the procedure into motion. We get $x_1 = (1/2)(1.5 + 2/1.5) = 1.4166667$, $x_2 = (1/2)(1.4166667 + 2/1.4166667) = 1.4142157$, $x_3 = 1.4142136$, and so on. After only three steps, we get the value of $\sqrt{2}$ correct to eight places. Newton's method (named after ISAAC NEWTON) is particularly easy to implement on a programmable calculator. Note, however, that the method fails if at some stage $f'(x_n) = 0$. In that case, a different initial guess x_0 might remedy the situation.

normal line A line perpendicular to another line.
> *See also* LINE(S), EQUATIONS OF, PERPENDICULAR; PERPENDICULAR LINES.

***n*th derivative** *See* DERIVATIVE.

***n*th order differential equation** *See* ORDER OF A DIFFERENTIAL EQUATION.

***n*th partial sum** *See* INFINITE SERIES, PARTIAL SUM OF.

numerical integration A computational procedure in which we replace the definition of a definite integral as a limit of a Riemann sum by a Riemann sum without going to the limit. There are several ways of doing this.
> *See* MIDPOINT RULE; SIMPSON'S RULE; TRAPEZOID RULE. *See also* INTEGRAL, DEFINITE; RIEMANN SUM.

odd function A function f with the property $f(-x) = -f(x)$ for all x in the domain of the function. Examples are all odd-powered functions ($f(x) = x^n$ with n an odd integer), the trigonometric functions sin x, tan x, and csc x, and the rational function $x/(1 - x^2)$. The graph of an odd function has an S-shaped symmetry about the origin; that is, the graph will look the same when turned through $180°$ about the origin.
> *See also* EVEN FUNCTION.

one-sided limit *See* LIMIT, ONE-SIDED.

one-to-one function A function f with the property that if $x_1 \neq x_2$, then $f(x_1) \neq f(x_2)$; that is, two different values of x always produce two different values of y (or stated differently, $f(x_1) = f(x_2)$ implies $x_1 = x_2$). For example, the function $f(x) = x^3$ is one-to-one, but the function $f(x) = x^2$ is not, because $f(-3) = f(3) = 9$. The graph of a one-to-one function satisfies the *horizontal line test*. Moreover, f has an inverse if and only if f is one-to-one and onto.
> *See also* ONTO FUNCTION.

onto function This concept applies only if the range of a function f(x) is thought of as a pre-assigned set R to which f "maps" the x-values of the domain D. In this case, a function is said to be "onto" if every number in R is the image of some number in D; that is, if no number in R is "left out." For example, the function $y = f(x) = x^2$ with range $[0, \infty)$ is onto, because every nonnegative number is the square of some real number; but the function $y = f(x) = x^2$ with range $(-\infty, \infty)$ is not onto, because a negative number is not the square of any real number.
> *See also* FUNCTION, DEFINITION OF; RANGE; HORIZONTAL LINE TEST.

open interval *See* INTERVAL, OPEN.

optimization The process of finding the maximum or minimum values of a function. Usually the word is used in connection with some applied "real-life" problem.

order of a differential equation The highest-order derivative of the unknown function y appearing in the equation. For example, the equation $xy' + y^2 = \ln x$ is of order 1, because the highest-order derivative of y is the first.
    See also DIFFERENTIAL EQUATION.

ordered pair A pair of numbers, written (x, y). Note that this is different from the pair (y, x). The pair can be interpreted as the coordinates of a point in a two-dimensional coordinate system.

ordinate The second number of an ordered pair, usually called the y-coordinate.

origin The point of intersection of the x- and y-axes in a two-dimensional rectangular coordinate system. Its coordinates are $(0, 0)$.

orthogonal trajectories Two families of curves such that each curve of one family intersects every curve of the other family at 90°. For example, the family of concentric circles with center at the origin and the family of straight lines through the origin form orthogonal trajectories.

parabola The set of all points p in the plane whose distance from a fixed point, called the *focus,* and from a fixed line, called the *directrix,* is always the same (of course, the focus should not lie on the directrix). The point midway between the focus and the directrix is called the *vertex*—the point where the parabola makes its sharpest turn. The line passing through the focus and perpendicular to the directrix is the *axis*—the line of symmetry of the parabola. If the vertex is at the origin, the focus on the y-axis at $(0, p)$, and the directrix the horizontal line $y = -p$, where $p > 0$, then the equation of the parabola is $x^2 = 4py$. We often rewrite this equation in the form $y = ax^2$, where $a = 1/4p$; when $a = 1$, it is called the *standard* parabola—the simplest of all second-degree polynomials. Similarly, the equation $x = y^2$ represents a parabola with a vertical directrix and horizontal axis.
    Latus rectum: The chord passing through the focus of the parabola and perpendicular to its axis. It intersects the parabola $x^2 = 4py$ at the points $(-2p, p)$ and $(2p, p)$ and has length $4p$.
    Polar equation of: In the following, let $p > 0$:

parabola

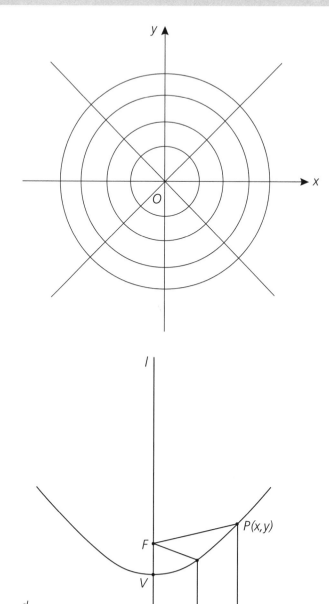

Parabola (F = focus;
V = vertex; d = directrix;
l = axis)

A parabola with horizontal directrix y = p and focus at the origin:
$$r = p/(1 + \sin \theta).$$
A parabola with horizontal directrix y = –p and focus at the origin:
$$r = p/(1 - \sin \theta).$$
A parabola with vertical directrix x = p and focus at the origin:
$$r = p/(1 + \cos \theta).$$
A parabola with vertical directrix x = –p and focus at the origin:
$$r = p/(1 - \cos \theta).$$

Reflective property of: If we imagine the parabola to be coated with a reflective surface, then all rays of light falling on it in a direction parallel to the axis are reflected to the focus ("focus" in Latin means "fireplace"). This property is used in TV dish antennas, which have a parabolic cross section with the detector at the focus, where it collects signals coming from satellites in orbit.

Standard equation of: The equation $y - k = a(x - h)^2$, representing a parabola with a vertical axis and vertex at (h, k). This equation can be rewritten as $y = ax^2 + bx + c$, where *a, b,* and *c* are constants and $a \neq 0$. Depending on whether a > 0 or a < 0, the parabola opens up or down. Its *Y*-intercept is (0, c). Similarly, the equation $x - k = a(y - h)^2$ represents a parabola with a horizontal axis and vertex at (k, h).

See also QUADRATIC FUNCTION.

parallel lines Lines having the same slope (if the lines are vertical, their slope is undefined).

parametric equations A pair of equations in which *x* and *y* are each expressed as a function of a third variable, called the *parameter.* The two equations together describe a curve parametrically. For example, the pair of equations x = r cos θ, y = r sin θ describe a circle with center at the origin and radius *r.* In this case we can eliminate the parameter θ between the two equations by squaring each and adding; we get $x^2 + y^2 = r^2$, which is the rectangular equation of the same circle. Often, however, such an elimination is difficult or impossible to carry out, in which case we have no choice but to describe the curve parametrically; an example is the *cycloid,* whose parametric equations are x = a(θ – sin θ), y = a(1 – cos θ).

partial fractions, decomposition into A technique used when finding an indefinite integral (antiderivative) of a rational function whose denominator is a product of several factors. We split the expression into a sum of individual fractions, each with a single factor in the denominator. For example, the fraction $1/(x^2 - 1) = 1/[(x + 1)(x - 1)]$ can be split into the *partial fractions* –1/[2(x + 1)] and 1/[2(x – 1)] (it is easy to check that their sum is equal to the original expression); these partial fractions are easier to integrate than the original expression. For rules describing how the decomposition should be done, consult any calculus textbook.

particular solution of a differential equation A solution that does not involve arbitrary constants. For example, a particular solution of the equation $y'' + y = 0$ is $y = 2 \cos x + 3 \sin x$. A particular solution is obtained from the *general solution* by taking into account the *initial conditions.* In the example given above, the particular solution is obtained from the general solution $y = A \cos x + B \sin x$ by imposing on it the initial conditions $y(0) = 2$, $y'(0) = 3$.

See also GENERAL SOLUTION OF A DIFFERENTIAL EQUATION.

partition The division of an interval into a number of subintervals (not necessarily of equal length). This is usually done when setting up a RIEMANN SUM.

percentage error See ERROR, RELATIVE.

period See PERIODIC FUNCTION.

periodic function A function f with the property $f(x + P) = f(x)$ for every number x in the domain of f. The smallest number P for which this is true is called the *period.* For example, the function $\sin x$ has a period 2π, because 2π is the smallest number for which $\sin(x + 2\pi) = \sin x$ for all x (note that the equation $\sin(x + 4\pi) = \sin x$ is also true for all x, but the period is still 2π). Periodic functions are important in the study of oscillations and waves.

perpendicular lines Two lines that intersect at 90°. Generally, the slopes of perpendicular lines are negative reciprocals of each other; that is, if one is m, the other is $-1/m$. This does not apply when one line is horizontal and the other vertical, because the slope of a vertical line is undefined.

piecewise-defined function A function having different definitions for different intervals in its domain. Such a function requires more than one formula for its definition, but it is still considered *one* function. An example is the function $f(x) = \{-x$ for $x \leq 0$ and x^2 for $x > 0\}$, whose graph consists of two parts—the straight line $y = -x$ for $x \leq 0$, and the parabola $y = x^2$ for $x > 0$. Because the two parts meet at the same point $(0, 0)$, this function is continuous. The function $g(x) = \{-1$ for $x \leq 0$ and 1 for $x > 0\}$, on the other hand, is *discontinuous* at $x = 0$.

See also DISCONTINUITY.

point   On the number line: The word is used synonymously with the x-coordinate of the point.

In the plane: The word is used synonymously with the x- and y-coordinates of the point, that is, the ordered pair (x, y).

point of inflection *See* INFLECTION POINT.

point-slope form *See* LINE(S), EQUATIONS OF, POINT-SLOPE FORM.

polar coordinates A coordinate system in which a point in the plane is located in terms of its distance r from the origin and the angle θ between the positive x-axis and the line connecting the point to the origin, measured counterclockwise in radians. The polar coordinates of a point with rectangular coordinates (x, y) are written as (r, θ). For example, the point with rectangular coordinates (0, 1) has polar coordinates (1, $\pi/2$), and the point with rectangular coordinates (–1, 1) has polar coordinates ($\sqrt{2}$, $3\pi/4$). To convert from rectangular to polar coordinates, use the formulas $r = \sqrt{x^2 + y^2}$, $\theta = \tan^{-1} y/x$. To convert from polar to rectangular coordinates, use the formulas $x = r \cos \theta$, $y = r \sin \theta$.

polar equation of a curve An equation of the form $r = f(\theta)$, where r and θ are the polar coordinates of a point on the curve, and f is a given

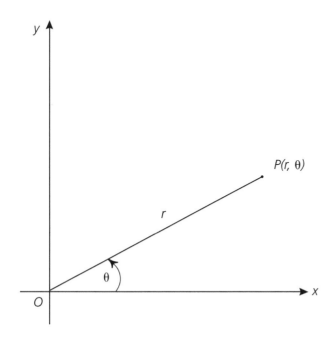

Polar coordinates

function. For example, the equation $r = e^{a\theta}$, where a is a constant, describes a logarithmic spiral, a curve often found in nature. A nautilus shell has the form of a logarithmic spiral, as does the arrangement of seeds in a sunflower.

See also LOGARITHMIC SPIRAL; POLAR COORDINATES.

polar form of a complex number A complex number $x + iy$, where $i = \sqrt{-1}$ and x and y are expressed by their polar coordinates. Using the formulas $x = r \cos \theta$, $y = r \sin \theta$, we can write the number as $r(\cos \theta + i \sin \theta)$, where $r = \sqrt{x^2 + y^2}$ and $\theta = \tan^{-1} y/x$. For example, the complex number $1 + 2i$ has the polar form $r(\cos \theta + i \sin \theta)$, where $r = \sqrt{1^2 + 2^2} = \sqrt{5}$, $\theta = \tan^{-1} 2/1 \approx 63.4° \approx 1.107$ radians.

See also COMPLEX NUMBER; POLAR COORDINATES.

polynomial functions A family of functions whose general equation is $f(x) = a_n x^n + a_{n-1} x^{n-1} + \ldots + a_2 x^2 + a_1 x + a_0$. The coefficients a_i are real numbers, with $a_n \neq 0$; a_n is the *leading coefficient,* a_0 the *constant,* and n the *degree* of the polynomial. Polynomials are studied in algebra as well as in calculus, where they are regarded as functions of x. The domain of all polynomial functions is all real numbers.

Graphs of: The graphs of all polynomial functions are continuous. Their Y-intercept is at $(0, f(0)) = (0, a_0)$. The graph of a polynomial function of degree n crosses the x-axis at most n times, that is, it has at most n X-intercepts; it has at most $n - 1$ extremum points (points where the function has a relative maximum or relative minimum); and it has at most $n - 2$ inflection points (points where the graph changes from concave up to concave down, or vice versa). For example, the functions $f(x) = x^3 - x$, $g(x) = x^3 + x$, and $h(x) = x^2 + 1$ have respectively 3, 1 and 0 X-intercepts, 2, 0, and 1 extremum points, and 1, 1, and 0 inflection points.

The behavior of the graph of a polynomial function for $x \to \infty$ and $x \to -\infty$ depends on whether the degree n is even or odd and on whether the leading coefficient a_n is positive or negative:

n = even, $a_n > 0$: the graph rises as $x \to \pm\infty$
n = even, $a_n < 0$: the graph falls as $x \to \pm\infty$
n = odd, $a_n > 0$: the graph rises as $x \to \infty$ and falls as $x \to -\infty$
n = odd, $a_n < 0$: the graph falls as $x \to \infty$ and rises as $x \to -\infty$

For algebraic properties of polynomial functions, consult any college algebra textbook.

Power Rule For differentiation: The rule $d(x^n)/dx = nx^{n-1}$. The rule applies to any n, whether integer, rational or irrational, or even non-real (though this last case is beyond ordinary calculus). For example $d(x^5)/dx = 5x^4$, $d(\sqrt{x})/dx = d(x^{1/2})/dx = (1/2)x^{-1/2} = 1/(2\sqrt{x})$, and $d(x^{\sqrt{2}})/dx = \sqrt{2}x^{\sqrt{2}-1}$.

Generalized: The rule $d(u^n)/dx = nu^{n-1}du/dx$, where $u = g(x)$. This is a result of combining the Power Rule and the Chain Rule. For example, $d(\sin^5 x)/dx = 5\sin^4 x \cdot (\cos x)$, the factor $\cos x$ coming from the derivative of $u = \sin x$.

See also CHAIN RULE.

For integration: The rule $\int x^n\, dx = x^{n+1}/(n+1) + C$, valid for all n except $n = -1$; C is the constant of integration. It is the inverse of the Power Rule for differentiation. For example, $\int x^5\, dx = x^6/6 + C$. For the case $n = -1$, we have $\int x^{-1}\, dx = \int (1/x)\, dx = \ln |x| + C$, where $\ln |x|$ is the natural logarithm of $|x|$; hence this case is not covered under the Power Rule.

power series An infinite series of the form $a_0 + a_1x + a_2x^2 + \ldots + a_nx^n + \ldots$, also written as $\sum_{i=0}^{\infty} a_i x^i$. Such a series converges in an open interval $(-R, R)$ centered at the origin; this interval is called the *interval of convergence* of the series, and the nonnegative number R is the *radius of convergence* (the word "radius" comes from the fact that if the real variable x is replaced by the complex variable $z = x + iy$, the series will converge for all z inside an open circle with radius R). If $R = 0$, the series *diverges,* while if $R = \infty$, it converges for all real numbers. At either endpoint of the interval of convergence, the series may or may not converge.

For example, the series $1 + x + x^2 + x^3 + \ldots$ converges for all x in the open interval $(-1, 1)$, but not at the endpoints; its sum inside the interval of convergence is $1/(1-x)$. The series $x - x^2/2 + x^3/3 - x^4/4 + - \ldots$ converges for all x in the half-open interval $(-1, 1]$, and its sum there is $\ln(1+x)$; for $x = 1$ we get the remarkable result $1 - 1/2 + 1/3 - 1/4 + - \ldots = \ln 2$. The series $1 + x + x^2/2! + x^3/3! + \ldots$ converges for all real x, and its sum is the exponential function e^x. Every power series inside its interval of converges defines a function $f(x)$; we say that $f(x)$ is *represented* by its power series.

Generalization of: An infinite series of the form $a_0 + a_1(x - c) + a_2(x - c)^2 + \ldots + a_n(x - c)^n + \ldots = \sum_{i=0}^{\infty} a_i(x - c)^i$. The constant c is called the *center* of the series. The series converges for all x in the open interval $(c - R, c + R)$ centered at $x = c$; as before, the nonnegative number R is called the radius of convergence.

Term-by-term differentiation and integration of: Inside its interval of convergence, a power series can be differentiated and integrated *term-by-term;* and the result is, respectively, the derivative and the indefinite integral of f(x). For example, the derivative of the series $1 + x + x^2 + \ldots$ on the interval $(-1,1)$ is the series $1 + 2x + 3x^2 + \ldots$; this series converges to the function $d/dx[1/(1-x)] = 1/(1-x)^2$. The indefinite integral (antiderivative) of the series $1 + x + x^2 + \ldots$ on the interval $(-1,1)$ is the series $x + x^2/2 + x^3/3 + \ldots$, which converges to the function $\int 1/(1-x)\,dx = -\ln(1-x)$.

See also MACLAURIN SERIES; TAYLOR SERIES; TAYLOR THEOREM.

present value The amount of money that must be deposited at present in order for it to grow to a specified future value, or balance, after the elapse of a specified time period. If the desired future value is A, the annual interest rate is r (expressed as a decimal), and the money is compounded n times a year for t years, then the present value, or principal, P, is given by the formula $P = A/(1 + r/n)^{nt}$ (or equivalently, $A(1 + r/n)^{-nt}$). For example, if it is desired to have a balance $200 in an account that pays an annual interest rate of 5% compounded four times a year (quarterly) after 10 years, one must deposit the amount $P = 200/(1 + 0.05/4)^{40} = \$121.68$. If the bank is using *continuous* compounding, the formula is $P = A/e^{rt}$ (or equivalently, Ae^{-rt}), where $e \approx 2.78$ is the base of natural logarithms. For the data given above, the present value for continuous compounding will be $P = 200/e^{0.05 \cdot 10} = \$121.31$.

See also FUTURE VALUE.

product rule The differentiation rule for a product of two functions f and g: $d/dx\,[f(x)g(x)] = f(x)\,d/dx\,g(x) + g(x)\,d/dx\,f(x)$, or in abbreviated notation, $(fg)' = fg' + gf'$ (of course, the order of performing these operations is immaterial: we can also write $(fg)' = f'g + g'f$). For example, $d/dx\,(x \sin x) = x \cdot (\cos x) + \sin x \cdot (1) = x \cos x + \sin x$.

product-to-sum identities *See* Section Four A.

p-series The infinite series $1/1^p + 1/2^p + 1/3^p + \ldots = \sum_{n=1}^{\infty} 1/n^p$, where p is a real number. This series converges for $p > 1$ and diverges for $p \leq 1$. If $p = 1$, the series is the HARMONIC SERIES.

Alternating: The series $1/1^p - 1/2^p + 1/3^p - + \ldots = \sum_{n=1}^{\infty} (-1)^{n+1}/n^p$. This series converges for all $p > 0$. If $p = 1$, the series is $1/1 - 1/2 + 1/3 - + \ldots$, which converges to the natural logarithm of 2.

Pythagorean identities *See* Section Four A.

Quadrant

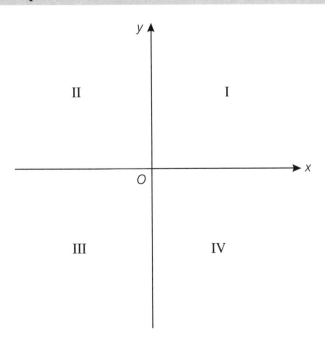

quadrant A region of the plane lying between two adjacent halves of the coordinate axes. There are four quadrants, counted counterclockwise from the positive *x*-axis and denoted by Roman numerals: I, II, III, and IV. The signs of *x* and *y* in the four quadrants are as follows:

Quadrant I: x > 0, y > 0
Quadrant II: x < 0, y > 0
Quadrant III: x < 0, y < 0
Quadrant IV: x > 0, y < 0

quadratic function A second-degree polynomial function, that is, a function of the form $f(x) = ax^2 + bx + c$, where *a*, *b*, and *c* are constants (coefficients) and $a \neq 0$. The graph of every quadratic function is a parabola with a vertical axis of symmetry. Depending on whether a > 0 or a < 0, the parabola opens up or down. The *Y*-intercept is at (0, c), and the two *X*-intercepts are the solutions of the quadratic equation $ax^2 + bx + c = 0$. If these solutions are non-real, the parabola does not cross the *x*-axis.

Standard form of: The equation $f(x) = a(x - h)^2 + k$, where the numbers *h* and *k* are the *x*- and *y*-coordinates of the vertex (the highest or lowest point of the parabola). Every quadratic function $f(x) = ax^2 + bx + c$ can be brought into standard form by completing the square; this leads to the formulas $h = -b/2a$, $k = f(-b/2a) = (4ac - b^2)/4a$. For example, the quadratic function $f(x) = 2x^2 + 3x + 1$ has the standard form $f(x) = 2(x + 3/4)^2 - 1/8$, showing that the vertex is at $(-3/4, -1/8)$. *See also* PARABOLA.

quotient of two functions The quotient, or ratio, of two functions f and g, written f/g. That is to say, $(f/g)(x) = f(x)/g(x)$ provided $g(x) \neq 0$. For example, if $f(x) = 2x + 1$ and $g(x) = 3x - 2$, then $(f/g)(x) = (2x + 1)/(3x - 2)$, provided $x \neq 2/3$.

Quotient Rule The differentiation rule for the quotient (ratio) of two functions f and g: $d/dx\ [f(x)/g(x)] = [g(x)\ d/dx\ f(x) - f(x)\ d/dx\ g(x)]/[g(x)]^2$, or in abbreviated notation, $(f/g)' = (gf' - fg')/g^2$ (note that unlike the Product Rule, here the order of operations does matter). For example, $d/dx\ [(\sin x)/x] = [x \cdot (\cos x) - (\sin x) \cdot 1]/x^2 = (x \cos x - \sin x)/x^2$. Of course, the rule is invalid when $g(x) = 0$.

radian An angular measure used in calculus and higher mathematics. In a circle of radius r, one radian is the angle, measured at the center of the circle, that subtends an arc length equal to r on the circumference. Because the circumference of a circle is $2\pi r$, and each arc length r corresponds to a central angle of one radian, we have 2π radians = 360°, or 1 radian = $360°/2\pi \approx 360°/6.283 = 57.296°$. Often, however, we express radian units as multiples of π, using the relation

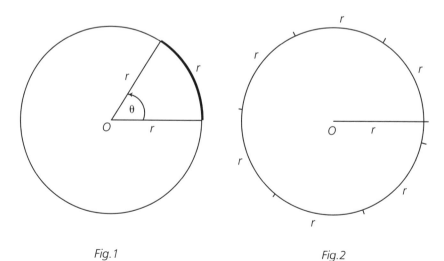

Radian: (1) 1 radian ≈ 57.296°; (2) 2π radians = 360° = one circumference

Fig.1   *Fig.2*

π radians = 180°. This gives us the following conversion table, where the designation "radians" has been dropped:

$$15° = \pi/12$$
$$30° = \pi/6$$
$$45° = \pi/4$$
$$60° = \pi/3$$
$$90° = \pi/2$$
$$180° = \pi$$
$$270° = 3\pi/2$$
$$360° = 2\pi$$

This table is limited to selected "special angles," but it can be extended to other special angles, for example, $135° = 90° + 45° = \pi/2 + \pi/4 = 3\pi/4$. To convert any angle from degree to radian measure, multiply by $\pi/180$ (approximately 0.017); to convert from radian to degree measure, multiply by $180/\pi$ (approximately 57.296).

radical function The family of functions $f(x) = \sqrt[n]{x}$, where n is a positive integer. The domain of a radical function depends on whether n is odd or even: if n is odd, the domain is all real numbers; if n is even, the domain is all nonnegative real numbers. A radical function can always be written as $f(x) = x^{1/n}$, so it is often regarded as a power function.

radius of convergence *See* POWER SERIES.

radius of curvature *See* CURVATURE.

range There is some disagreement as to the definition of range. Some authors define range as the set of all "images" of a function; that is, the set of all y-values of $y = f(x)$ as the independent variable x takes on all values of the domain. For example, the range of $y = x^2$ is all nonnegative real numbers, that is, all $y \geq 0$. Others define it as a preassigned set R to which the function f assigns, or "maps," the independent variable x as it covers the domain of f. In this definition, not all numbers in R are necessarily images of f. For example, we can define the function $y = x^2$ as a mapping from the set of all real numbers (the domain) to the set of all real numbers (the range), even though only nonnegative values of y are actually obtained. *See also* DOMAIN; FUNCTION, DEFINITION OF; ONTO FUNCTION.

rate of change Average: The ratio $[f(b) - f(a)]/(b - a)$, where $y = f(x)$ is a given function and a and b are two points in its domain. Also called *difference quotient* or *rise-to-run ratio* and denoted by $\Delta y/\Delta x$, where Δx and Δy are the increments in x and y, respectively. Geometrically,

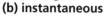

Rate of change:
(a) average;
(b) instantaneous

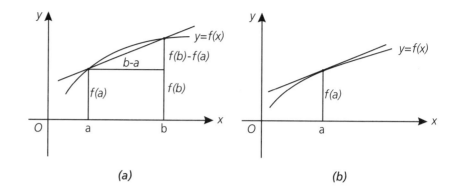

(a)   (b)

it is the slope of the *secant line* connecting the points (a, f(a)) and (b, f(b)). *See* DIFFERENCE QUOTIENT; SECANT LINE.

Instantaneous: The rate of change of a function y = f(x) at a point x = a of its domain. It is equal to the derivative of f(x) with respect to *x*, evaluated at the point x = a; that is, d/dx f(x) when x = a, or simply f′(a). Geometrically, it is the slope of the *tangent line* to the graph of f(x) at x = a.

See also DERIVATIVE; TANGENT LINE.

ratio of two functions *See* QUOTIENT OF TWO FUNCTIONS.

Ratio Test The infinite series $\sum_{i=1}^{\infty} a_i$ converges absolutely if $\lim_{i \to \infty} |a_{i+1}/a_i| < 1$, and diverges if $\lim_{i \to \infty} |a_{i+1}/a_i| > 1$. If $\lim_{i \to \infty} |a_{i+1}/a_i| = 1$, the test is inconclusive. For example, for the series $\sum_{i=1}^{\infty} (-2)^i/i! = -2/1 + 2^2/2! - 2^3/3! + - \dots$ we have $\lim_{i \to \infty} |a_{i+1}/a_i| = \lim_{i \to \infty} |[(-2)^{i+1}/(i+1)!]/[(-2)^i/i!]| = \lim_{i \to \infty} 2/(i+1) = 0$, so the series converges absolutely (that is, not only does the given series converge, but so does the same series with all terms positive). *See also* CONVERGENCE, ABSOLUTE; Section Four D.

rational function An expression of the form h(x) = f(x)/g(x), where f(x) and g(x) are polynomial functions. The domain of h(x) consists of all *x* values for which g(x) ≠ 0. If the ratio f(x)/g(x) is in lowest terms, then the graph of h(x) has a vertical asymptote whenever g(x) = 0. At all other points h(x) is continuous. If h(x) tends to a limit *c* as x → ∞ or x → −∞, the line y = c is a horizontal asymptote of the graph of h(x). For example, the function h(x) = (2x + 1)/(x² + 2x − 3) has vertical asymptotes at x = 1 and x = –3 (because the denominator $x^2 + 2x - 3 = (x - 1)(x + 3)$ is zero at these points) and a horizontal asymptote y = 0 (because $\lim_{x \to \infty} h(x) = 0$).

rational number   A real number that can be written as a ratio of two integers. Examples are 3/2, –4/5, 7 (because 7 = 7/1), and 0. The decimal expansion of a rational number is either terminating or nonterminating and repeating. Examples are 3/2 = 1.5 and 4/33 = 0.121212 . . .
   See also IRRATIONAL NUMBER.

rationalizing   A process that removes a radical (or several radicals) from a numerator or a denominator of a quotient. This is done by multiplying and dividing by the *conjugate* of the expression that contains the radical (or radicals) to be removed. For example, to

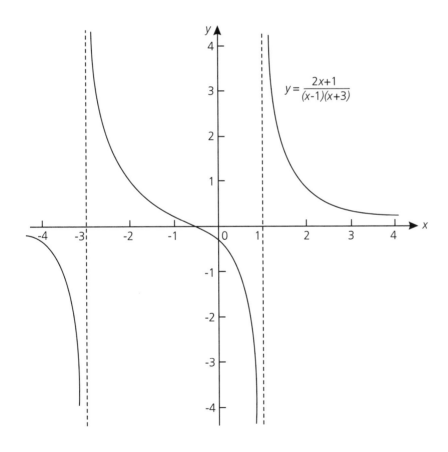

$$y = \frac{2x+1}{(x-1)(x+3)}$$

Rational function

remove the radical from the denominator of $2/(3 + \sqrt{5})$, we multiply and divide the quotient by $3 - \sqrt{5}$: $[2/(3 + \sqrt{5})] \cdot [(3 - \sqrt{5})/(3 - \sqrt{5})] = 2(3 - \sqrt{5})/[(3 + \sqrt{5})(3 - \sqrt{5})] = 2(3 - \sqrt{5})/[3^2 - (\sqrt{5})^2] = 2(3 - \sqrt{5})/(9 - 5) = (3 - \sqrt{5})/2$. As another example, to remove the radical from the numerator of $(\sqrt{2} - \sqrt{3})/5$ we multiply and divide by $\sqrt{2} + \sqrt{3}$ and get $-1/[5(\sqrt{2} + \sqrt{3})]$.

Rationalizing is used in calculus when trying to find the limits of various expressions. For example, in order to find the derivative of $\sqrt{x}$ from first principles (that is, from the definition of derivative), we need to find the limit of $[\sqrt{x + h} - \sqrt{x}]/h$ as $h \to 0$. If we try to substitute $h = 0$ into the quotient, we end up with the indeterminate expression $0/0$. To go around this, we rationalize the numerator; this transforms the quotient into $1/[\sqrt{x + h} + \sqrt{x}]$, whose limit as $x \to 0$ is $1/(2\sqrt{x})$.

real number Any decimal. The real numbers are represented by the points on the number line and form a *number continuum*.

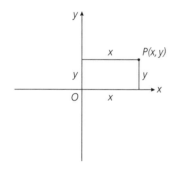

Rectangular coordinates

rectangular coordinates The set of ordered pairs (x, y), where x and y refer to a two-dimensional rectangular coordinate system. The number x is the distance of a point P from the y-axis, and the number y is its distance from the x-axis. Also known as *Cartesian coordinates* after their inventor, RENÉ DESCARTES.

reference angle For a given angle θ in standard position in a rectangular coordinate system (measured counterclockwise from the positive x-axis), the reference angle θ' is found from the following table, where θ is in radians:

Terminal side of θ in Quadrant I: $\theta' = \theta$
Terminal side of θ in Quadrant II: $\theta' = \pi - \theta$
Terminal side of θ in Quadrant III: $\theta' = \theta - \pi$
Terminal side of θ in Quadrant IV: $\theta' = 2\pi - \theta$

reflection of a graph Let the graph have the equation $y = f(x)$. Then:

Reflection in the x-axis: the new graph has the equation $y = -f(x)$.
Reflection in the y-axis: the new graph has the equation $y = f(-x)$.
Reflection in the line $y = x$: the new graph has the equation $y = f^{-1}(x)$, where f^{-1} is the inverse of the function f.

reflective property

Of an ellipse, *See* ELLIPSE.
Of a parabola, *See* PARABOLA.

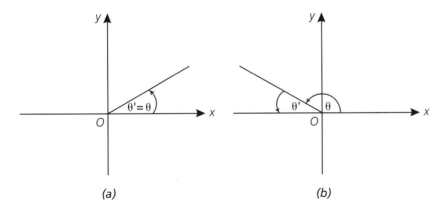

(a)   (b)

Reference angles:
(a) Quadrant I: $\theta' = \theta$;
(b) Quadrant II: $\theta' = \pi - \theta$;
(c) Quadrant III: $\theta' = \theta - \pi$;
(d) Quadrant IV: $\theta' = 2\pi - \theta$

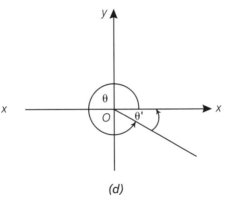

(c)   (d)

related rates When two variables x and y are related to each other through an equation, while each is a function of a third variable t (for example, time), the rates of change of x and y with respect to t are also related to each other. To find the relation between the two rates of change, we use *implicit differentiation.* As an example, suppose a ladder of length 5 feet is leaning against a wall. The lower end is dragged away from the wall at a rate of 2 feet per second. At what rate is the upper end sliding down the wall when the lower end is 3 feet from the wall? Denoting the distance of the lower end of the ladder from the wall by x and the distance of the upper end from the floor by y and using the Pythagorean Theorem, we have $x^2 + y^2 = 25$. We now differentiate both sides of this equation with respect to time t, bearing in mind that x and y change

Related rates

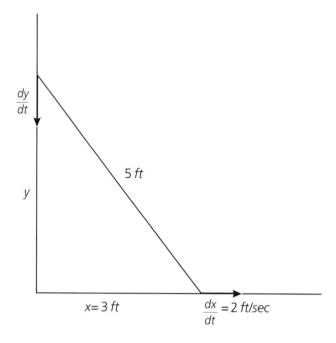

continuously with t; we get $2x(dx/dt) + 2y(dy/dt) = 0$, or, after canceling 2, $x(dx/dt) + y(dy/dt) = 0$. Now when x = 3, we have $y = \sqrt{25 - 9} = \sqrt{16} = 4$. Putting x = 3, y = 4, and dx/dt = 2 into the related rates equation, we get $3 \cdot 2 + 4(dy/dt) = 0$, from which we find dy/dt = − 6/4 = −1.5 feet per second (the minus sign shows that the upper end moves down while the lower end is moving away from the wall).

See also IMPLICIT DIFFERENTIATION.

relative error *See* ERROR, RELATIVE.

relative maximum *See* MAXIMUM, RELATIVE.

relative minimum *See* MINIMUM, RELATIVE.

remainder (of a Taylor polynomial) *See* TAYLOR THEOREM.

removable discontinuity *See* DISCONTINUITY, REMOVABLE.

Riccati differential equation The first-order differential equation $y' = p(x)y^2 + q(x)y + r(x)$, where p, q, and r are given functions of x. Named after Jacopo Riccati (1676–1754).

Richter scale A logarithmic scale that measures the magnitude of earthquakes. The magnitude is given by the formula $M = \log(I/I_0)$, where I is the intensity of the earthquake (in some appropriate units of energy), I_0 is some standard or reference intensity, and "log" means common (base 10) logarithm. Because the Richter scale only measures the magnitude of one earthquake relative to that of another, the actual value of I_0 is immaterial. A tenfold increase in the intensity translates into an increase of one unit on the Richter scale.

As an example, the 1989 California earthquake measured 7.1 on the Richter scale, while the earthquake that devastated San Francisco in 1906 measured 8.4. If we denote the intensities of these earthquakes by I_1 and I_2, respectively, we have $7.1 = \log (I_1/I_0)$, $8.4 = \log (I_2/I_0)$. Subtracting and using the properties of logarithm, we have $1.3 = \log (I_2/I_0) - \log (I_1/I_0) = \log [(I_2/I_0)/(I_1/I_0)] = \log (I_2/I_1)$, so $I_2/I_1 = 10^{1.3} \approx 19.95$. Thus the 1906 earthquake was nearly 20 times more powerful than that of 1989.

See also LOGARITHM.

Riemann sum Given a continuous function f(x) on a closed interval [a,b], we divide [a, b] into n subintervals (not necessarily of equal length) by the points $x_0 = a, x_1, x_2, \ldots, x_n = b$. The ith subinterval is $[x_{i-1}, x_i]$, with length $\Delta x_i = x_i - x_{i-1}$. In each subinterval we select one

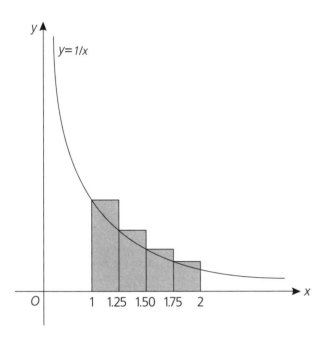

Riemann sum

representative point and denote it by x_i^*, i = 1, 2, . . ., n. We now write down the value of f(x) at each of these points, that is, $f(x_i^*)$ for i = 1, 2, . . ., n, and then form the product $f(x_i^*)\Delta x_i$. Finally, we add up all these products, that is, we form the sum $f(x_1^*)\Delta x_1 + f(x_2^*)\Delta x_2 + . . . + f(x_n^*)\Delta x_n$, or in the *sigma notation*, $\sum_{i=1}^{n} f(x_i^*)\Delta x_i$. This last expression is called a *Riemann sum* of f(x) over [a, b], named after GEORG FRIEDRICH BERNHARD RIEMANN. Note that we used the phrase "*a* Riemann sum," not "*the* Riemann sum," because the value of the sum depends not only on f(x) and on [a, b], but also on the particular division of [a, b] into subintervals, as well as on our choice of the x_i^*s. If f(x) ≥ 0 over the entire interval [a, b], we can interpret a Riemann sum as the sum of the areas of n rectangles, the *i*th of which has a base of length Δx_i and height $f(x_i^*)$.

If we divide [a, b] into *n equal* subintervals, they have the same length $\Delta x = (b – a)/n$ (note that we have dropped the subscript *i* under the Δx). Then the Riemann sum becomes $\sum_{i=1}^{n} f(x_i^*)\Delta x$, or simply $\Delta x \sum_{i=1}^{n} f(x_i^*)$, because Δx is constant and can be moved outside the sigma sign. This usually simplifies the calculation of the sum.

As an example, let us find the Riemann sum of f(x) = 1/x over [1, 2], using n = 4 equal subintervals. We then have $\Delta x = (2 – 1)/4 = 1/4 = 0.25$, and the points of subdivision are $x_0 = 1$, $x_1 = 1.25$, $x_2 = 1.50$, $x_3 = 1.75$, $x_4 = 2$. For x_i^* we choose the *left* endpoint of each subinterval; that is, $x_i^* = x_{i-1}$. The Riemann sum then becomes (1/4)[1/1 + 1/1.25 + 1/1.50 + 1/1.75] = 0.7595 (rounded to four places). Had we chosen the *right* endpoint of each subinterval (that is, $x_i^* = x_i$), the sum would be (1/4)[1/1.25 + 1/1.50 + 1/1.75 + 1/2] = 0.6345. Each of these numbers is an estimate of the actual area under the graph of y = 1/x from x = 1 to x = 2; the first sum gives us an overestimation, while the second is an underestimation. To find the exact area, we need to calculate the *definite integral* of 1/x from x = 1 to x = 2 (which turns out to be ln 2, or 0.6931). Indeed, the Riemann sum is the key to defining the definite integral.

See also INTEGRAL, DEFINITE.

right-handed limit *See* LIMIT, ONE-SIDED.

rise-to-run ratio *See* DIFFERENCE QUOTIENT; RATE OF CHANGE, AVERAGE.

Rolle's Theorem Let *f* be a continuous function on a closed interval [a, b] and differentiable on the open interval (a, b). If f(a) = f(b), there exists at least one number *c* in (a, b) for which f′(c) = 0 (that is, the tangent line

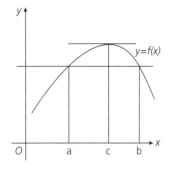

Rolle's Theorem

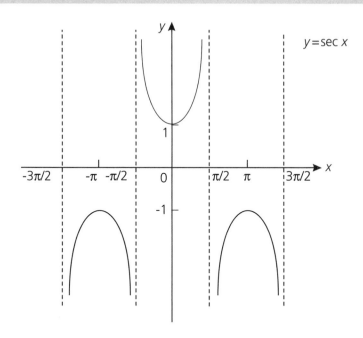

Secant function

to the graph of f at $x = c$ is horizontal). The theorem is named after MICHEL ROLLE and is mainly used to prove the *Mean Value Theorem.*

Root Test The infinite series $\sum_{i=1}^{n} a_i$ converges absolutely if $\lim_{i \to \infty} \sqrt[i]{|a_i|} < 1$, and diverges if $\lim_{i \to \infty} \sqrt[i]{|a_i|} > 1$. If $\lim_{i \to \infty} \sqrt[i]{|a_i|} = 1$, the test is inconclusive. For

example, for the series $\sum_{i=1}^{n} (1/i)^i$ we have $\lim_{i \to \infty} \sqrt[i]{\left|\left(\frac{1}{i}\right)^i\right|} = \lim_{i \to \infty} \frac{1}{i} = 0$, so the series converges absolutely.

    See also CONVERGENCE, ABSOLUTE; Section Four D.

secant function The function $y = \sec x = 1/\cos x$. Its domain is all real numbers except $\pm\pi/2, \pm 3\pi/2, \ldots$, and its range is $(-\infty, -1] \cup [1, \infty)$. It is a periodic function with period 2π. Its derivative is $d/dx \sec x = \sin x/\cos^2 x = \tan x \sec x$. For other properties of the secant function, *See* Section Four A.

secant line A straight line connecting two points on the graph of a function $y = f(x)$. If the points are $(a, f(a))$ and $(b, f(b))$, then the slope of the secant line is $m = [f(b) - f(a)]/(b - a)$, and its point-slope equation is $y - f(a) = m(x - a)$, or equivalently $y - f(b) = m(x - b)$.

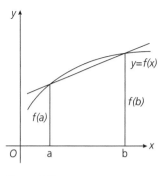

Secant line

second derivative The derivative of the first derivative of a function $y = f(x)$. We write d^2y/dx^2 (derived from $d/dx(dy/dx)$), or $d^2/dx^2[f(x)]$. Another notation is $f''(x)$, or simply y''. For example, if $y = x^5$, then $y'' = d^2/dx^2(x^5) = d/dx(5x^4) = 20x^3$. The second derivative measures the rate of change of the slope of a graph. In physics, it is the *acceleration* of an object when its position is known as a function of time.

Second Derivative Test Let f be a twice-differentiable function on an open interval containing a point c, and let $f'(c) = 0$ (that is, c is a critical point of f). Then:

If $f''(c) > 0$, f has a relative minimum at $x = c$.
If $f''(c) < 0$, f has a relative maximum at $x = c$.
If $f''(c) = 0$, the test is inconclusive (in this case, use the First Derivative Test).

Note that unlike the First Derivative Test, here we only consider a critical point c where $f'(c) = 0$, not where $f'(c)$ is undefined; in the latter case, $f''(c)$ obviously is not defined either. As an example, let $f(x) = 2x^3 - 9x^2 - 24x + 5$. We have $f'(x) = 6x^2 - 18x - 24 = 6(x^2 - 3x - 4) = 6(x + 1)(x - 4)$, so $f'(x) = 0$ when $x = -1$ and 4. Now $f''(x) = 12x - 18$, so $f''(-1) = 12 \cdot (-1) - 18 = -30 < 0$ and f has a relative maximum at $x = -1$. Similarly, $f''(4) = 12 \cdot (4) - 18 = 30 > 0$, so f has a relative minimum at $x = 4$. As another example, consider $f(x) = x^4$. We have $f'(x) = 4x^3$, so $f'(0) = 0$. Also, $f''(x) = 12x^2$, so $f''(0) = 0$ and the Second Derivative Test is inconclusive. But from the shape of the graph of $f(x)$ (a parabola-like graph, though more flat around $x = 0$), or by using the First Derivative Test, we know that f has a relative (indeed, an absolute) minimum there. On the other hand, the function $f(x) = x^3$, for which again $f'(0) = f''(0) = 0$, has neither a relative minimum nor a relative maximum at $x = 0$, but an inflection point there.

See also CRITICAL NUMBER (VALUE, POINT); FIRST DERIVATIVE TEST.

separable differential equation A differential equation of the form $y' = f(x)/g(y)$, where f and g are given functions of x and y, respectively. Writing $y' = dy/dx$ and regarding dx and dy as differentials, we can write the equation in the equivalent form $f(x)\,dx = g(y)dy$, in which the two variables are "separated" (each side containing only one variable). Integrating both sides, we get the general solution in implicit form, $\int f(x)\,dx = \int g(y)\,dy + C$, where C is the constant of integration.

See also DIFFERENTIAL.

sequence An ordered set of numbers. A sequence may be finite or infinite. If finite, we write it as $a_1, a_2, \ldots, a_n$; if infinite, as $a_1, a_2, \ldots, a_n, \ldots$ Although any set of ordered numbers is a sequence, there is usually some rule that tells us how to obtain the ith term a_i when the index i is given. For example, the ith term of the sequence 1, 1/2, 1/4, 1/8, $\ldots$ (a geometric progression) is $1/2^{i-1}$ (note that the exponent is $i-1$ rather than i because the first term is $1 = 1/2^0$), so we can find, say, the 10th term: $1/2^9 = 1/512$. Sometimes the rule gives a_i in terms of a_{i-1}, in which case it is called a *recursion formula.* For example, the terms of the Fibonacci sequence 1, 1, 2, 3, 5, 8, 13, 21, $\ldots$ are given by the rule $a_1 = a_2 = 1$ and $a_i = a_{i-2} + a_{i-1}$ for $i = 3, 4, 5, \ldots$

Bounded: A sequence whose terms never get larger than a number M (called an *upper bound*), or never get smaller than a number N (a *lower bound*). For example, the geometric progression 1, 1/2, 1/4, 1/8, $\ldots$ has an upper bound 1 and a lower bound 0. Of course, any number larger than M is also an upper bound, and any number smaller than N is also a lower bound, so upper and lower bounds are not unique. But the *smallest* upper bound is unique, as is the *largest* lower bound; either may or may not be a member of the sequence. In the example given, the smallest upper bound is 1 (a member of the sequence), and the largest lower bound is 0 (not a member).

Convergence of, *See* CONVERGENCE OF A SEQUENCE.
Divergence of, *See* DIVERGENCE OF A SEQUENCE.
Limit of, *See* LIMIT OF A SEQUENCE.
Monotone, *See* MONOTONIC, SEQUENCE.

series The sum of the terms of a sequence. A series may be finite or infinite; if finite, we write it as $a_1 + a_2 + \ldots + a_n$, or, using the sigma notation, $\sum_{i=1}^{n} a_i$; if infinite, as $a_1 + a_2 + \ldots + a_n + \ldots = \sum_{i=1}^{\infty} a_i$. An infinite series may or may not converge. For example, the geometric series $1 + 1/2 + 1/4 + 1/8 + \ldots$ converges to the limit 2, while the harmonic series $1 + 1/2 + 1/3 + 1/4 + \ldots$ diverges. If an infinite series converges to a limit S, we say that the series *has the sum S* and write $\sum_{i=1}^{\infty} a_i = S$.

Absolute convergence of, *See* CONVERGENCE, ABSOLUTE.
Alternating: A series whose terms alternate between positive and negative. For example, the harmonic series with alternating signs is $1 - 1/2 + 1/3 - 1/4 + - \ldots$, whose sum is ln 2. *See* CONVERGENCE, CONDITIONAL.
Conditional convergence of, *See* CONVERGENCE, CONDITIONAL.
Divergence of, *See* DIVERGENCE OF A SERIES.

Geometric, *See* GEOMETRIC SERIES.
Harmonic, *See* HARMONIC SERIES.
Limit of, *See* LIMIT OF A SERIES.
Maclaurin, *See* MACLAURIN SERIES.
Power, *See* POWER SERIES.
Sequence of partial sums of: The sequence a_1, $(a_1 + a_2)$, $(a_1 + a_2 + a_3)$, ...
Taylor, *See* TAYLOR SERIES.
Telescopic, *See* TELESCOPIC SERIES.

shell method A method for finding the volume of a solid of revolution. We imagine the solid to be made up of infinitely many thin concentric shells, or cylinders, whose common axis is the axis of revolution. The total volume is found by integrating the volumes of these shells over the length of the solid. If the solid is generated by revolving the graph of y = f(x) about the y-axis, the volume is

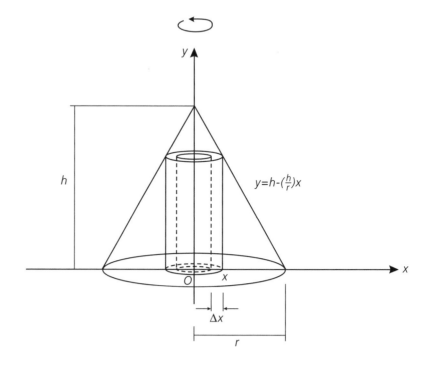

Shell method

given by the formula $V = 2\pi \int_a^b yx\,dx = 2\pi \int_a^b f(x)x\,dx$, where a and b are the upper and lower limits of the interval in question. (Note: this formula applies to a solid revolving about the y-axis. If the solid revolves about the x-axis, the formula is $V = 2\pi \int_c^d xy\,dy$, where now $x = g(y)$ and the integration is with respect to y). As an example, a circular cone of base radius r and height h can be generated by revolving the line $y = h - (h/r)x$ about the y-axis. Its volume is given by $V = 2\pi \int_0^r [h - (h/r)x]x\,dx$. After expanding the integrand and integrating each term, we get $V = \pi r^2 h/3$.

See also DISK METHOD; SOLID OF REVOLUTION.

shift of a graph Let $y = f(x)$ be a function whose graph is known, and let c be a positive number. Then:

The graph of $f(x - c)$ is identical with the graph of $f(x)$ but shifted c units to the right.

The graph of $f(x + c)$ is identical with the graph of $f(x)$ but shifted c units to the left.

The graph of $f(x) + c$ is identical with the graph of $f(x)$ but shifted c units up.

The graph of $f(x) - c$ is identical with the graph of $f(x)$ but shifted c units down.

For example, the graph of $y = (x - 1)^2$ is a parabola identical to the parabola $y = x^2$ but shifted one unit to the right. The graph of $y = x^2 + 1$ is identical to the parabola $y = x^2$ but shifted one unit up. The graph of $y = (x - 1)^2 + 1$ is identical to the parabola $y = x^2$ but shifted one unit to the right *and* one unit up. Note that the rules given above apply for *positive c*. If c is negative, the direction of the shift reverses.

See also TRANSFORMATION OF A GRAPH.

sigma notation An abbreviated notation for the sum of the terms of a finite or infinite sequence. For the finite sum $a_1 + a_2 + \ldots + a_n$ we write $\sum_{i=1}^{n} a_i$; for the infinite sum $a_1 + a_2 + \ldots + a_n + \ldots$ we write $\sum_{i=1}^{\infty} a_i$ (in the latter case, the sum is defined only if the series converges). The numbers 1 and n provide the range of summation and must be given. The subscript i is the *index of summation;* it merely serves as a "counter" and can be replaced by any other letter not already in use; that is, $\sum_{i=1}^{n} a_i = \sum_{j=1}^{n} a_j = \sum_{k=1}^{n} a_k$, and so on. The sigma notation has the following properties, where for brevity we drop the range:

$\sum ca_i = c\sum a_i$ (a constant can be moved outside the $\sum$)
$\sum(a_i \pm b_i) = \sum a_i \pm \sum b_i$ (the sigma of a sum or difference is equal to the sum or difference of the sigmas, respectively)

In addition, $\sum_{i=1}^{n} c = nc$; that is, the sum of a constant added n times to itself is n times the constant.
   See also SUMMATION FORMULAS.

simple harmonic motion (SHM) A to-and-fro motion of a point described by the equation $y = a \sin(bt + c)$. Here $|a|$ is the *amplitude* (the maximum deviation to either side of the equilibrium point $y = 0$), b is the *angular frequency,* and c is the *phase.* The *period* is $T = 2\pi/b$, and the *frequency* is $f = 1/T = b/2\pi$. The equation $y = a \cos(bt + c)$ also describes SHM. Many physical systems approximately follow SHM, as for example a vibrating spring or the oscillations of a swing. Electrical phenomena often follow a similar law, such as the voltage in an alternating current.

simple interest A financial procedure by which a bank pays interest on the original investment (the principal) only. If the principal is denoted by P, the annual interest rate by r, and the money is compounded once a year, then the balance A after t years is given by the formula $A = P(1 + rt)$. For example, if $P = \$100$ and $r = 5\% = 0.05$, the balance after 10 years will be $A = 100(1 + 0.05 \cdot 10) = \$150$.
   See also COMPOUND INTEREST.

Simpson's Rule A procedure for approximating the value of a definite integral: $\int_a^b f(x)\, dx \approx [f(x_0) + 4f(x_1) + 2f(x_2) + 4f(x_3) + \ldots + 2f(x_{n-2}) + 4f(x_{n-1}) + f(x_n)]\Delta x/3$, where x_i, i = 1, 2, . . ., n, are the points of subdivision of the interval [a, b] into n equal subintervals, each of length $\Delta x = (b - a)/n$. (Note: here n must be an even integer.) The procedure is named after THOMAS SIMPSON.
   See also DEFINITE INTEGRAL; MIDPOINT RULE; RIEMANN SUM; TRAPEZOID RULE.

simulation In a narrow sense, this word is synonymous with MODELING. *Computer* simulation, in which a real-life problem is translated into a series of computer-generated codes, is rapidly becoming an indispensable tool in every branch of science.

sine function The function $y = f(x) = \sin x$. Its domain is all real numbers, and its range the interval [–1, 1]. Its graph is periodic—it repeats every

Sine function

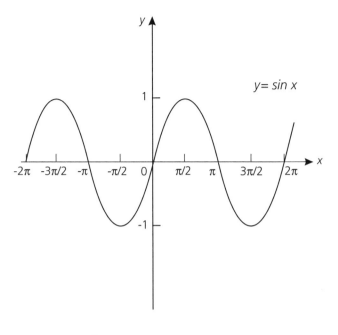

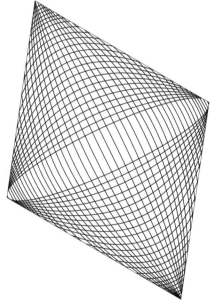

Superposition of two sine oscillations at right angles

2π radians. The graph of sin x is identical in shape to that of cos x, but shifted relative to it by one quarter of a cycle, that is, by $\pi/2$ (*See* COSINE FUNCTION). In applications, especially in the study of vibrations and waves, the vertical distance from the x-axis to either the highest or the lowest point of the graph (that is, 1) is called the *amplitude,* and the period 2π is the *wavelength.* More generally, the function $y = a \sin bx$ has amplitude $|a|$ and period $2\pi/b$. One can also shift the graph left and right; this is represented by the function $y = a \sin (bx + c)$. The derivative of the sine function is $d/dx \sin x = \cos x$.

See Section Four A for other properties of the sine function.

slant asymptote *See* ASYMPTOTE, SLANT.

slope The steepness of a straight line, or, in case of a non-straight line graph, the steepness of the tangent line to the graph at a point x = a. In the former case, we denote the slope by m and find it from the formula $m = (y_2 - y_1)/(x_2 - x_1)$, where (x_1, y_1) and (x_2, y_2) are two points on the line. (If the line is vertical, we have $x_1 = x_2$, and the slope is undefined.) In the latter case, the slope is the value of the derivative of f(x) at x = a; that is d/dx f(x) at x = a, or $f'(a)$. *See also* DERIVATIVE; DIFFERENCE QUOTIENT; RATE OF CHANGE.

slope-intercept form *See* LINE(S), EQUATIONS OF, SLOPE-INTERCEPT FORM.

smooth curve Loosely speaking, a curve that has no breaks or sharp corners. More technically, the graph of a continuously differentiable function. All polynomials have smooth graphs, as do the rational, trigonometric, and exponential functions in their respective domains. An example of a non-smooth curve is the graph of the absolute-value function $y = |x|$, which has a corner at x = 0.

solid of revolution A solid generated by revolving a two-dimensional curve about a fixed line (usually the *x*- or *y*-axis). For example, if we revolve the parabola $y = x^2$ about the *y*-axis, we get a *paraboloid of revolution.* Revolving the same parabola about the *x*-axis produces a different solid.

See DISK METHOD; SHELL METHOD for methods of finding the volume of a solid of revolution.

square-root function The function $y = f(x) = \sqrt{x}$. Its domain is all nonnegative real numbers, that is, $x \geq 0$ (or in interval notation, $[0, \infty)$), and its range is the same interval along the y-axis, that is, $y \geq 0$. Its graph is the upper half of the horizontal parabola $x = y^2$. Its derivative is

Solid of revolution

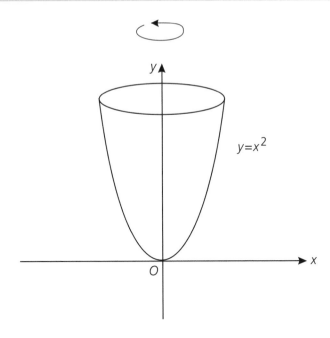

$y=x^2$

d/dx $\sqrt{x}$ = 1/(2$\sqrt{x}$). It is often convenient to write the square-root function as a power function y = $x^{1/2}$. Its derivative is then $x^{-1/2}$/2.

Squeeze Theorem Let *f*, *g*, and *h* be three functions such that f(x) ≤ g(x) ≤ h(x) for all *x* in an open interval containing a point *c*, except possibly at x = c. If $\lim_{x\to c}$ f(x) = $\lim_{x\to c}$ h(x) = L, then $\lim_{x\to c}$ g(x) exists and is also equal to L. That is, the graph of g(x) is "squeezed" between those of f(x) and h(x) near x = c.

substitution, method of A method of finding certain indefinite integrals (antiderivatives), when the integrand (the function inside the integral sign) is of the "right form." The key to a successful application of the method is to write the integrand (including the symbol dx) as the *differential* of a known function F(u); that is, to transform the original integral into an integral of the form ∫f(u) du, where f(u)du = dF(u), or equivalently, f(u) = d/du F(u) = F′(u), so that F(u) is an antiderivative of f(u); we then have ∫f(u) du = F(u) + C. For example, the integral ∫2sin 2x dx can be written as ∫sin u du, where u = 2x and du = 2dx; we then have ∫sin u du = −cos u + C = −cos 2x + C (the last step is

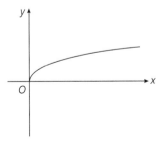

Square root function

Squeeze Theorem

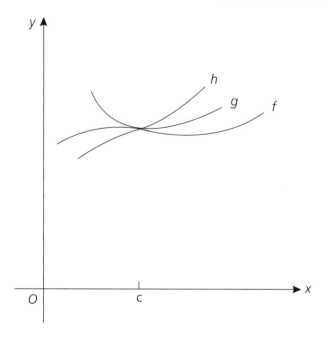

necessary because we want to express the answer in terms of *x,* not *u*). The general formula is:

$$\int f(g(x))g'(x)\,dx = \int f(u)\,du = F(u) + C = F(g(x)) + C,$$
$$\text{where } u = g(x) \text{ and } F'(u) = f(u).$$

As another example, consider $\int [2x/(1 - x^2)]\,dx$. Substituting $u = 1 - x^2$ and $du = -2x\,dx$, the integral is transformed into $-\int du/u = -\ln|u| + C = -\ln|1 - x^2| + C$. Not every integral lends itself to this method, and it takes some experience to find the right substitution that will "do the trick." In a sense, the method is the reverse of the *Chain Rule.*

See also DIFFERENTIAL.

sum rule For differentiation: $d/dx\,[f(x) \pm g(x)] = d/dx\,f(x) \pm d/dx\,g(x)$, or in abbreviated notation, $(f \pm g)' = f' \pm g'$. In words: the derivative of a sum or difference of two functions is the sum or difference of their derivatives, respectively. For example, $d/dx\,(2x^3 + 3x^4) = d/dx\,(2x^3) + d/dx\,(3x^4) = 6x^2 + 12x^3$.

For integration: $\int [f(x) \pm g(x)]\,dx = \int f(x)\,dx \pm \int g(x)\,dx$. In words: the indefinite integral (antiderivative) of a sum or difference of two functions is the sum or difference of their indefinite integrals, respectively. A corresponding rule applies to definite integrals. As an example, $\int(2x^3 + 3x^4)\,dx = \int 2x^3\,dx + \int 3x^4\,dx = x^4/2 + 3x^5/5 + C$ (here we combined the constants of integration of the two integrals into a single constant).

sum-to-product identities *See* Section Four A.

summation formulas Formulas for finding the sums of powers of integers. The most commonly used are:

Sum of the first n integers: $\displaystyle\sum_{i=1}^{n} i = 1 + 2 + 3 + \ldots + n = n(n + 1)/2$

Sum of the first n squares:

$$\sum_{i=1}^{n} i^2 = 1^2 + 2^2 + 3^2 + \ldots + n^2 = n(n + 1)(2n + 1)/6$$

Sum of the first n cubes: $\displaystyle\sum_{i=1}^{n} i^3 = 1^3 + 2^3 + 3^3 + \ldots + n^3 = [n(n + 1)/2]^2$

Although these formulas are purely algebraic, they are often used in calculus when finding the Riemann sum of various power functions.

summation notation *See* SIGMA NOTATION.

surface of revolution *See* SOLID OF REVOLUTION.

symmetry A term widely used in different branches of mathematics and science (and in art as well). In calculus, the term refers to the graph of a function $y = f(x)$ in relation to the coordinate axes:

If $f(-x) = -f(x)$ for all x-values in the domain of f, the graph has a symmetry with respect to the origin.
If $f(-x) = f(x)$ for all x-values in the domain of f, the graph has a symmetry with respect to the y-axis.

In the former case, f is called an *odd function;* in the latter case, an *even function.* For example, the graph of $y = x^3$ has a symmetry with respect to the origin, while that of $y = x^2$ has a symmetry with respect to the y-axis. We can also talk of the symmetry of a graph with respect to the x-axis, although in this case the graph does not represent a function (the vertical line test fails). Another way to look at these symmetries is as follows:

If the equation describing the graph does not change when x is replaced by $-x$, the graph is symmetric with respect to the y-axis.
If the equation describing the graph does not change when y is replaced by $-y$, the graph is symmetric with respect to the x-axis.
If the equation describing the graph does not change when x is replaced by $-x$ and y by $-y$, the graph is symmetric with respect to the origin.

Here the phrase "the equation does not change" means that the new equation is equivalent to the original. For example if we replace x by $-x$ and y by $-y$ in the equation $y = 2x - 3x^3$, we get the equation $-y = 2(-x) - 3(-x)^3$, which is equivalent to the original equation (after simplifying, the right-hand side becomes $-2x + 3x^3$); thus the graph described by $y = 2x - 3x^3$ is symmetric with respect to the origin. We should stress that a graph may have other symmetries not related to the axes; for example, every parabola is symmetric with respect to its own axis (*See* PARABOLA), but not necessarily with respect to the x- or y-axes.

table of integrals *See* Section Four C.

tangent function The function $y = \tan x = \sin x/\cos x$ (sometimes the notation tg x is used). Its domain is all real numbers except $\pm\pi/2$, $\pm3\pi/2$, $\pm5\pi/2$, ..., that is, except the odd multiples of $\pi/2$. At these points, the graph of tan x has vertical asymptotes and is discontinuous there. Otherwise the graph is continuous and crosses the x-axis at $\pm\pi$, $\pm2\pi$, $\pm3\pi$, ..., that is, at the points midway between adjacent asymptotes. The range of tan x is all real numbers, which means that every real number is the tangent of some value of x. The tangent function is periodic with

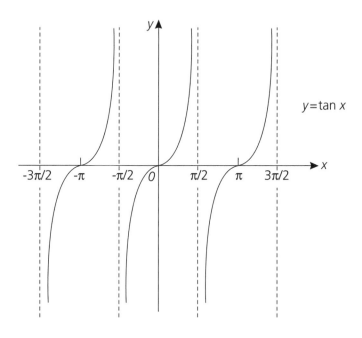

Tangent function

period π (unlike the sine and cosine functions, which have period 2π). The derivative is $d/dx \tan x = 1/\cos^2 x = \sec^2 x$.

See also COTANGENT FUNCTION; Section Four A for additional properties of the tangent function.

tangent line Loosely speaking, a line that touches a curve locally but does not cross it (we say "locally," because a tangent line can touch a curve at one point but cross it at another point). The importance of the tangent line in calculus comes from the fact that near the point of tangency, the graph of a function can be approximated by its tangent line, whose slope is the value of the derivative of the function at the point of tangency.

See also DERIVATIVE; LINEAR APPROXIMATION.

tangent line approximation *See* LINEAR APPROXIMATION.

Taylor polynomial Let f be a function that can be differentiated n times at a point x = c. The nth Taylor polynomial of f at x = c is the polynomial

$$P_n(x) = f(c) + f'(c)(x - c) + f''(c)(x - c)^2/2! + f'''(c)(x - c)^3/3! + \ldots + f^{(n)}(c)(x - c)^n/n!$$

Here n! (read: "n factorial") is the product $1 \cdot 2 \cdot 3 \cdot \ldots \cdot n$, and $f^{(n)}(c)$ is the nth derivative of f at x = c (note that the "zeroth" derivative of f is f itself, so the index n starts with 0).

As an example, let $f(x) = \ln x$ and c = 1. We have $f(1) = \ln 1 = 0$, $f'(1) = 1/x|_{x=1} = 1$, $f''(1) = -1/x^2|_{x=1} = -1$, $f'''(1) = 2/x^3|_{x=1} = 2$, and so on. The Taylor polynomial for f(x) at c = 1 is thus $(x-1) - (x-1)^2/2 + (x-1)^3/3 - (x-1)^4/4 + - \ldots + (-1)^{n+1}(x-1)^n/n$. The polynomial is named after BROOK TAYLOR. The special case when c = 0 is called the Maclaurin polynomial (after COLIN MACLAURIN). For example, the Maclaurin polynomial of sin x is $x - x^3/3! + x^5/5! - x^7/7! + - \ldots + (-1)^{n-1} x^{2n-1}/(2n-1)!$.

The nth Taylor polynomial of f can be used to approximate the values of f near a point x = c in its domain. For n = 1, the Taylor polynomial is $P_1(x) = f(c) + f'(c)(x - c)$, which is the linear approximation of f at x = c (*See* LINEAR APPROXIMATION). For n = 2, we get the second-degree approximation $P_2(x) = f(c) + f'(c)(x - c) + f''(c)(x - c)^2/2!$, and so on. Generally, the approximation will improve with each additional term.

See also TAYLOR SERIES; TAYLOR THEOREM.

Taylor series Let f be a function that can be differentiated infinitely many times at a point x = c. The power series

$$f(c) + f'(c)(x - c) + f''(c)(x - c)^2/2! + f'''(c)(x - c)^3/3! + \ldots$$

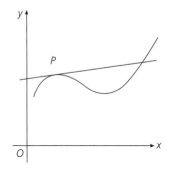

Tangent line

is called the *Taylor series of f at x = c* (named after BROOK TAYLOR). Note that in contrast with the Taylor polynomial, the Taylor series is an *infinite* series. For example, the Taylor series for the function f(x) $= \ln x$ at x = 1 is $(x - 1) - (x - 1)^2/2 + (x - 1)^3/3 - (x - 1)^4/4 + - \ldots$. The special case of the Taylor series when c = 0 is called the Maclaurin series, named after COLIN MACLAURIN. For example, the Maclaurin series of sin x is $x - x^3/3! + x^5/5! - x^7/7! + - \ldots + (-1)^{n-1} x^{2n-1}/(2n - 1)! + \ldots$. The question of whether the Taylor series converges to f(x) is discussed under TAYLOR THEOREM.

See also MACLAURIN SERIES.

Taylor Theorem Let *f* be a function that can be differentiated n + 1 times on an interval *I* that contains the point x = c. Then f(x) can be written as: $f(x) = f(c) + f'(c)(x - c) + f''(c)(x - c)^2/2! + f'''(c)(x - c)^3/3! + \ldots + f^{(n)}(c)(x - c)^n/n! + R_n(x)$, where $R_n(x) = f^{(n+1)}(z)(x - c)^{n+1}/(n + 1)!$ for some point *z* in *I*. The term $R_n(x)$ is called the *remainder* of the Taylor expansion (specifically, the LaGrange form of the remainder; there are other forms). The question of whether the Taylor series for f(x) actually converges to f(x) is answered by the following theorem:

Let *f* be a function that can be differentiated infinitely many times in an open interval *I* containing the point x = c. Then $f(x) = f(c) + f'(c)(x - c) + f''(c)(x - c)^2/2! + \ldots + f^{(n)}(c)(x - c)^n/n! + \ldots$ if and only if $R_n(x)$ tends to 0 as $n \to \infty$; that is, if and only if $\lim_{n \to \infty} R_n(x) = 0$. In that case we say that f(x) is *represented* by its Taylor series. Corresponding statements hold for the Maclaurin series (the special case when c = 0). Most elementary functions are indeed represented by their Taylor or Maclaurin series, but the proof that $\lim_{n \to \infty} R_n(x) = 0$ in each case may be lengthy and is often omitted.

See also MACLAURIN SERIES; TAYLOR POLYNOMIAL; TAYLOR SERIES.

telescopic series A series in which all terms except the first and last cancel out in pairs: $(a_1 - a_2) + (a_2 - a_3) + (a_3 - a_4) + \ldots + (a_{n-1} - a_n)$. Its sum "collapses" to $a_1 - a_n$, from which the name "telescopic" comes. For example, the series $1/(1 \cdot 2) + 1/(2 \cdot 3) + 1/(3 \cdot 4) + \ldots + 1/[(n - 1)n] = (1/1 - 1/2) + (1/2 - 1/3) + (1/3 - 1/4) + \ldots + [1/(n - 1) - 1/n] = 1 - 1/n$. Since 1/n tends to zero as $n \to \infty$, the *infinite series* $1/(1 \cdot 2) + 1/(2 \cdot 3) + 1/(3 \cdot 4) + \ldots$ converges to the sum 1.

See also PARTIAL FRACTIONS.

total revenue function If the price of producing and selling *x* units of a commodity is *p* (in dollars), then the total revenue of selling those *x* units is px. As a rule, p depends on the number of units sold, so it is a function of x, called the *price function* and written p(x). Thus the total revenue is xp(x) and is itself a function of x.

transcendental functions The class of all non-algebraic functions; that is, all functions that cannot be expressed as a finite combination of the elementary functions (polynomials, ratios of polynomials, and radical functions). Examples are the trigonometric functions and their inverses, the exponential and logarithmic functions, and algebraic combinations of these. Also included are various "higher" functions studied in advanced mathematics.
See also ALGEBRAIC FUNCTIONS; ELEMENTARY FUNCTIONS.

transcendental number A number that is not algebraic (is not a zero of a polynomial function with integer coefficients). Examples are the numbers π and e.
See also ALGEBRAIC NUMBER.

transformation of a graph Let the graph be represented by the function y = f(x), and let c be a positive constant. Then:

y = f(x – c) shifts the graph c units to the right.
y = f(x + c) shifts the graph c units to the left.
y = f(x) + c shifts the graph c units up.
y = f(x) – c shifts the graph c units down.
y = –f(x) reflects the graph in the x-axis.
y = f(–x) reflects the graph in the y-axis.
y = $f^{-1}(x)$ reflects the graph in the line y = x.
y = cf(x) stretches the graph vertically in the ratio c:1 if c > 1, and shrinks it if 0 < c < 1.
y = f(cx) shrinks the graph horizontally in the ratio c:1 if c > 1, and stretches it if 0 < c < 1.

If c is *negative,* the direction of the shift in the first four cases reverses, while in the last two cases, the stretching or shrinking is accompanied by a reflection in the corresponding axis.
See also SHIFT OF A GRAPH.

translation of a graph *See* SHIFT OF A GRAPH.

trapezoid rule A procedure for approximating the value of a definite integral:
$$\int_a^b f(x)\, dx \approx \sum_{i=1}^{n} \{[f(x_{i-1}) + f(x_i)]/2\}\Delta x, \text{ where } x_i, i = 1, 2, \ldots, n, \text{ are}$$
the points of subdivision of the interval [a, b] into n equal subintervals, each of length $\Delta x = (b - a)/n$. (Note the difference

Triangle inequality

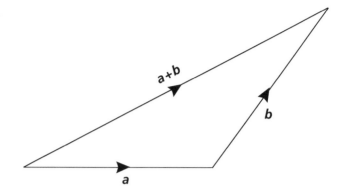

between this formula and the Midpoint Rule, despite their superficial similarity.)

See also DEFINITE INTEGRAL; MIDPOINT RULE; RIEMANN SUM; SIMPSON'S RULE.

Triangle Inequality The statement that the absolute value of the sum of two real numbers is never greater than the sum of their absolute values: $|a + b| \le |a| + |b|$, with equality if and only if a and b have the same sign. For example, $|2 + 3| = 5 = |2| + |3|$, $|2 + (-3)| = |-1| = 1 < |2| + |-3| = 5$, and $|(-2) + (-3)| = |-5| = 5 = |-2| + |-3|$. If we replace the numbers a and b with *vectors a* and *b,* the inequality $|\mathbf{a} + \mathbf{b}| \le |\mathbf{a}| + |\mathbf{b}|$ says that the length of a side in a triangle is never greater than the sum of the lengths of the other two sides, with equality if and only if *a* and *b* are along the same line.

trigonometric functions The set of six functions sin x, cos x, tan x, csc x, sec x, and cot x. *See also* each individual function; Section Four A.

trigonometric identities *See* Section Four A.

trigonometric substitutions A class of substitutions, or change of variables, used to find certain indefinite integrals (antiderivatives). In the following, let a > 0:

For integrals involving $\sqrt{a^2 - x^2}$, substitute x = a sin θ ($-\pi/2 \le \theta \le \pi/2$), dx = a cos θ dθ. This transforms the radical into a cos θ.

For integrals involving $\sqrt{x^2 - a^2}$, substitute x = a sec θ, dx = a sec θ tan θ dθ. This transforms the radical into ±a tan θ, with the "plus" sign if $0 \le \theta < \pi/2$ (that is, if x ≥ a) and the "minus" sign if $\pi/2 < \theta \le \pi$ (that is, x ≤ –a).

For integrals involving $\sqrt{a^2 + x^2}$, substitute $x = a \tan \theta$ $(-\pi/2 < \theta < \pi/2)$, $dx = \sec^2 \theta \, d\theta$. This transforms the radical into $a \sec \theta$.

As an example, let us find $\int dx/\sqrt{16 - x^2}$. We substitute $x = 4 \sin \theta$, $dx = 4 \cos \theta \, d\theta$. This transforms the integral into $\int 4 \cos \theta \, d\theta/(4 \cos \theta)$ $= \int d\theta = \theta + C = \arcsin x/4 + C$. It is convenient to memorize these substitutions with the aid of a right triangle.

unit circle The circle with radius 1 and center at the origin. Its equation in rectangular coordinates is $x^2 + y^2 = 1$, and in polar coordinates $r = 1$.

variable A quantity that can assume different values taken from a given set (the domain).

Dependent: The variable y in the expression $y = f(x)$ (or any equivalent letter in a similar expression, such as $w = g(u)$).

Independent: The variable x in the expression $y = f(x)$ (or any equivalent letter in a similar expression, such as $w = g(u)$).

Of integration: The variable with respect to which we compute an integral. This variable is recognized from the symbol dx inside the integral. For example, the variable of integration in $\int x \sin z \, dx$ is x, while z is regarded as constant; but if the integral is $\int x \sin z \, dz$, then the variable of integration is z, while x is regarded as constant. The two integrals are equal to $(x^2 \sin z)/2 + C$ and $-x \cos z + C$, respectively.

velocity The rate of change of the position of an object with respect to time. If a particle moves along the x-axis, its position is a function of time, $x = f(t)$, and its velocity is $v = dx/dt = f'(t)$.

vertex Of an ellipse, *See* ELLIPSE.
Of a hyperbola, *See* HYPERBOLA.
Of a parabola, *See* PARABOLA.

vertical asymptote *See* ASYMPTOTE, VERTICAL.

vertical line A line parallel to the y-axis. Its equation is $x = a$, where a is a constant, and its X-intercept is $(a, 0)$. For example, the equation $x = -3$ represents a vertical line through the point $(-3, 0)$. A vertical line has no slope, or more precisely, its slope is undefined; therefore, we cannot write its equation in the slope-intercept form.

vertical line test A graphical test that lets us see if a given graph is the graph of a function. If every vertical line intersects the graph at one point at most, the graph is that of a function; otherwise it is not (of course, a vertical line x = a may not intersect the graph at all; this happens if x = a is outside the domain of the function).

vertical shift of a graph Let the graph have the equation y = f(x). If we shift (translate) it *c* units up (where *c* is a positive number), its new equation is y = f(x) + c. If shifted *c* units down (where *c* is still positive), its equation becomes y = f(x) – c. For example, the graph of $y = x^2 - 1$ is identical to the graph of $y = x^2$ (a parabola) but shifted one unit down.

> *See also* HORIZONTAL SHIFT OF A GRAPH; TRANSFORMATION OF A GRAPH.

vertical tangent line A vertical line tangent to a curve at some point on it. For example, the line x = 1 is a vertical tangent line to the unit circle at the point (1, 0). Because the slope of a vertical line is undefined, the value of the derivative of the function describing the curve will also be undefined. In fact, one way to discover the existence of a vertical tangent line is to look for points where the function is defined but its derivative is not. For example, the graph of $y = \sqrt[3]{x} = x^{1/3}$ has a vertical tangent line at x = 0, because the derivative $y' = (1/3)x^{-2/3} = 1/(3x^{2/3})$ is undefined at x = 0.

viewing rectangle The region of the coordinate system displayed by a graphing calculator ("grapher"). The viewing rectangle can be adjusted to the particular graph under study; this is done by opening the window menu.

volume Loosely speaking, the amount of space that fills a closed surface. Except for a few simple surfaces (for example, a solid of revolution), finding the volume involves calculating a double integral and falls outside the domain of this book.

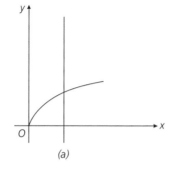
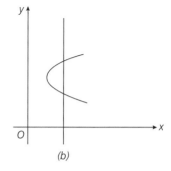

Vertical line test: (a) a function; (b) not a function

(a)

(b)

Vertical tangent line

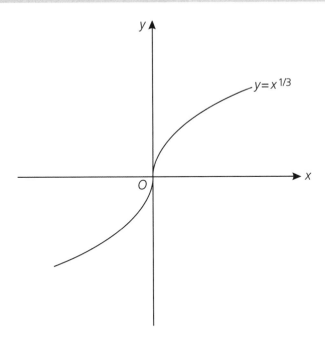

Wallis's product The infinite product $(2/1) \cdot (2/3) \cdot (4/3) \cdot (4/5) \cdot (6/5) \cdot (6/7) \cdot \ldots = \pi/2$. Named after its discoverer, JOHN WALLIS.

washer method *See* DISK METHOD.

work Let an object be acted on by a force F(x) that varies with the distance x along the x-axis. The work done by this force when moving the object from x = a to x = b is given by the integral $\int_a^b F(x)\, dx$. For example, when an elastic spring is stretched by an amount x from its neutral position, the force exerted on the spring is proportional to x (Hooke's Law); that is, F = kx. The work involved in this action is $\int_0^x kt\, dt = kx^2/2$ (note that we have denoted the variable of integration by t, so as not to confuse it with the upper limit x).

X-intercept *See* INTERCEPT, X-INTERCEPT.

Y-intercept *See* INTERCEPT, Y-INTERCEPT.

zero of a function A number c that is a solution of the equation f(x) = 0. For example, the zeros of the function sin x are the solutions of the equation sin x = 0, namely x = 0, ±π, ±2π, . . . Geometrically, the zeros of a function are the X-intercepts of its graph. When f(x) is a polynomial, its zeros are also called the *roots* of the polynomial (not to be confused with a square root). For example, the roots of the polynomial $x^3 - 4x^2 + x + 6$ are –1, 2, and 3. Except for linear and quadratic polynomials (or polynomials that can easily be factored), finding the zeros of a polynomial may not be easy; there are, however, numerical methods to find them approximately.

SECTION TWO
BIOGRAPHIES

Abel, Niels Henrik (1802–29) Norwegian algebraist, one of the founders of group theory, a branch of modern abstract algebra. The son of a Lutheran minister, he was poor during all of his short life, which in part prevented him from securing a position at any prestigious universities. He is perhaps best known for proving that the general *quintic equation* (a polynomial equation of degree 5) cannot be solved in terms of the elementary algebraic operations. The proof put an end to the search for a formula, however complicated, that would provide a solution to such an equation, similar to the quadratic formula for solving any quadratic (second-degree) equation. He also worked on a class of functions known as *elliptic functions,* a kind of generalization of the trigonometric functions that play a role in higher mathematics. Abel died of tuberculosis at the age of 26. Two days after his death, a letter arrived informing him of his appointment as professor of mathematics at the university of Berlin.

Agnesi, Maria Gaetana (1718–99) Italian mathematician, one of a handful of women before the 20th century who had chosen mathematics as a career. At a young age she was already fluent in several languages and versed in philosophy, physics and chemistry, botany and zoology, and mathematics. When only 14 she solved difficult problems in analytic geometry and mechanics. Her major work was the two-volume *Instituzioni analytiche ad uso della gioventu italiana* (Analytic institutions for the use of young Italians, 1748), in which she gave a complete presentation of algebra and calculus as it was then known. In 1750 she was appointed chair of mathematics and philosophy at the University of Bologna, but shortly thereafter she withdrew from academic life and devoted her remaining years to religious and social work. Ironically, her name is associated with a curve dubbed "the witch of Agnesi," although she had little to do with it; its equation is $y = 8a^3/(x^2 + 4a^2)$.

d'Alembert, Jean le Ronde (1717–83) French mathematician and physicist. At first he studied law and medicine, but later he turned to mathematics and made major contributions to the theory of partial differential equations (equations in which the unknown function depends on two or more independent variables, one of which is usually time). He applied his results to various branches of continuum mechanics—the motion of fluids, vibrations of strings, celestial mechanics, and the theory of tides. He formulated a principle in mechanics named after him, a generalization of Newton's third law of motion to moving bodies. In pure mathematics he clarified the limit concept—still a vague concept in his time—and paved the way to its modern definition. Together with the scholar Denis Diderot, he founded the great French *Encyclopédie* and served as its scientific editor until 1758.

Archimedes of Syracuse (ca. 287–212 B.C.E.) Greek mathematician, one of the greatest scientists of all time. Born in Syracuse on the island of Sicily (today in Italy), he was at heart a pure mathematician who was devoted to studying science for its own sake, but he also applied his discoveries to a wide range of practical problems. Among his discoveries are the law of the lever and the laws governing floating bodies. According to legend, when king Hieron suspected that his crown was not made of pure gold, he called upon Archimedes to investigate the matter. Archimedes immersed the crown in a bathtub, and from the amount of dispelled water he deduced that the crown was indeed a forgery; beside himself with excitement, he ran naked in the streets, shouting "eureka" (I found it!). In mathematics, he was the first to use the method of exhaustion, formulated by Eudoxus, to find the area under a segment of a parabola; in this he came tantalizingly close to discovering the integral calculus. He also showed that one can approximate the value of π to any degree of accuracy by "squeezing" a circle between a series of inscribed and circumscribing polygons of an ever larger number of sides; in this way he found that π was between $3\frac{10}{71}$ and $3\frac{10}{70}$, or, in decimal notation, between 3.14085 and 3.14286. Among his more practical inventions we mention the screw pump, still in use today, and a giant parabolic mirror, which according to legend he aimed at the Roman fleet besieging his city, setting it ablaze.

He was slain by a Roman soldier while sitting on the beach, drawing geometric figures in the sand. Some of Archimedes' writings survived, but many are lost. In 1906 one of these lost manuscripts was discovered in a monastery in Istanbul, giving us an invaluable glimpse into the mind of one of science's most legendary figures.

Aryabhata (475–ca. 550) A Hindu mathematician who is chiefly known through his work, the *Aryabhatiya,* in which he discusses arithmetic, trigonometry, the measurement of time, and astronomical tables. He shows how to enumerate numbers up to 100,000,000, gives a formula for the sum of an arithmetic progression, and states the quadratic formula (though in a form different from the one we learn in algebra today). The *Aryabhatiya* contains the first explicit reference to the sine function (although not by this name), and it gives a verbal rule for finding the numerical value of π, equivalent to the decimal value 3.1416.

Barrow, Isaac (1630–77) English mathematician and theologian who was Newton's predecessor as the Lucasian professor of mathematics at Cambridge. He later resigned his exalted position so that the young Newton, who had attended Barrow's lectures, could replace him. He anticipated many of the elements of the differential calculus but based his arguments mainly on geometry, failing to see the advantages of the analytic method.

Bernoulli, Jakob (James) (1654–1705) Swiss mathematician and the senior member of a remarkable dynasty of mathematicians that produced at least eight prominent members. Like most of the Bernoullis, he was born and lived in the quiet university town of Basel, on the banks of the Rhine river where the borders of Switzerland, France, and Germany meet. He and his brother Johann together studied the newly invented calculus and were among the first to apply it to numerous problems in mathematics and physics. Among their achievements was the solution to one of the outstanding problems of the time—to find the curve along which a particle under the force of gravity will slide down in the shortest possible time (this problem was known as the *brachistochrone,* from the Greek words meaning "shortest time"). The two brothers arrived at the solution independently and using different methods. The required curve turned out to be a *cycloid*—the curve traced by a point on the rim of a wheel as it rolls along a straight line. Rather than share in the glory of this discovery—which was submitted as an entry for a prestigious competition—the two started a bitter feud as to who should win the prize. This was typical of the Bernoullis, whose lives were colored by numerous rivalries, all in the name of science. The two brothers also did pioneering work on infinite series and on the theory of vibrating strings (another outstanding problem of the 18th century), and they studied the properties of various curves. Jakob's favorite was the *logarithmic spiral,* a graceful curve whose polar equation is $r = e^{a\theta}$. He asked that this spiral be engraved on his tombstone, a wish that was almost fulfilled: the engraver indeed cut a spiral on the headstone, but it was a *linear* instead of logarithmic spiral (the tombstone still stands at the central cathedral of Basel). Jakob also pioneered the field of mathematical probability; his book, *Ars conjectandi* (The Art of Conjecture, 1713), greatly influenced the development of this field. In this work Bernoulli discussed the theory of permutations and introduced the *Bernoulli numbers,* coefficients that appear in certain exponential series. He also coined the word *integral* for the antiderivative of a function.

Bernoulli, Johann (Jean) (1667–1748) Swiss mathematician, brother of Jakob (see above). He at first studied medicine but was soon drawn to mathematics. The two brothers did much of their work in the same field, often collaborating but also involving themselves in frequent rivalries. Johann made important contributions to continuum mechanics, particularly the theory of elasticity and fluid mechanics, and he wrote a classic treatise, *Hydraulica* (1738). This work, however, was immediately eclipsed by his son Daniel's book *Hydrodynamica,* published in the same year (it is in this work that the famous Bernoulli law governing the flight of an aircraft first appeared).

Bessel, Friedrich Wilhelm (1784–1846) German mathematician and astronomer. At the young age of 26 he became director of the newly opened Königsberg Observatory in Prussia (now Kaliningrad in Russia). There, in 1838, he achieved what no one had accomplished before—measuring the distance of a fixed star (the star was 61 Cygni, in the constellation Cygnus, the swan, some 11.1 light-years away). Thus for the first time the enormous dimensions of the universe beyond our solar system became known. Bessel also worked on the theory of gravitational perturbations, which led him to a new class of functions, the *Bessel functions.* These are solutions of the differential equation $x^2y'' + xy' + (x^2 - n^2)y = 0$, where $n \geq 0$ is a constant (not necessarily an integer). The nature of the solutions greatly depends on n. For $n = 1/2, 3/2, 5/2, \ldots$, they can be written in closed form in terms of the sine and cosine functions; otherwise they can only be expressed as infinite series. The Bessel equation shows up in many applications; for example, the vibrations of a circular membrane, such as that of a drumhead, are governed by Bessel's equation with $n = 0$. The solutions in this case bear certain similarities to the sine and cosine functions, but their zeros are not equally spaced.

Boole, George (1815–64) English mathematician who founded (with Augustus De Morgan but independently of him) symbolic logic. Boole was largely self-educated but nevertheless earned a professorship of mathematics at Queen's College at Cork, Ireland, in 1849. He was one of the first to treat the subject of *invariants*—quantities that do not change their value even if we change the coordinate system in which they are defined (an example is the discriminant $b^2 - 4ac$ of the quadratic expression $ax^2 + 2bxy + cy^2$, which is invariant under translations and rotations of the coordinates). This subject has become of great importance in modern algebra and mathematical physics. Boole also worked on what is called today the *finite difference* method—a numerical solution to a differential equation, in which the differentials dx and dy are replaced by the finite differences Δx and Δy. However, his interest gradually shifted to logic, and it is in this field that he is most known. His two books, *Analysis of Logic* (1847) and *Laws of Thought* (1854), became classics in the field. In these works he introduced algebraic operations that could be applied to propositional statements such as "if p then q" (written p $\Rightarrow$ q), "not p" (~p), "p or q" (p $\cup$ q), and so on. Boole thus transformed the subject into a computational topic, now known as *Boolean algebra,* where valid logical deductions could be made from given premises by a series of algebraic operations. Today Boolean algebra is used in designing logical circuits for computers; thus a once entirely theoretical subject has become one of enormous practical importance.

Briggs, Henry (1561–1630) English mathematician who was professor of mathematics at Oxford. His main claim for fame came from his visit in 1616 to John Napier, inventor of logarithms, at which Briggs convinced Napier to base his logarithms on base 10 (common logarithms). Napier was already too old to rework his tables, so Briggs himself undertook the task. His logarithms remained the basis of today's tables and were the principal computational aid available to scientists until the electronic handheld calculator appeared on the market in the 1970s. Logarithmic tables appeared in every college algebra textbook until the 1990s, when they were finally put to rest.

Cantor, Georg (1845–1918) German mathematician, founder of modern set theory. By looking at a set, and especially an *infinite* set, as a whole unit, Cantor was able to clarify the concept of mathematical infinity, which up until then had been shrouded in vagueness. Using the simple device of *one-to-one correspondence,* Cantor showed that an infinite set may have just as many elements as a *subset* of itself. For example, by pairing every natural number with its double, (that is, 1 is matched with 2, 2 with 4, 3 with 6, and so on), he showed that the set of natural numbers is just as numerous as the set of even numbers—defying the common notion that there are twice as many natural numbers as even numbers. Cantor called every set that can be put in a one-to-one correspondence with the natural numbers a *countable,* or *denumerable* set. These sets include the odd numbers, the squares, the primes, and, amazingly, even the rational numbers. But the set of real numbers cannot be put into a one-to-one correspondence with the natural numbers and is therefore non-denumerable: its infinity is of a higher order than that of denumerable sets.

Cantor went on to create an entire hierarchy of infinities, in which all sets with the same order of infinity can be matched one-to-one with each other but not with sets of a different order. Cantor's ideas were revolutionary for his time and at first were resisted by several conservative mathematicians, but they were gradually absorbed into mainstream mathematics and nowadays are taught in our elementary schools.

Cardan, Girolamo (**Cardano**; 1501–76) Italian scientist who, like most scientists of his time, spread his talents over many fields, including mathematics, physics, astronomy, astrology, and medicine. In the latter field he wrote the first work on typhus. As a mathematician, he is best known for his work *Ars magna*, in which he laid down a formula for solving cubic (third-degree) and quartic (fourth-degree) equations. Unfortunately, this got him embroiled in a bitter controversy with his contemporary Nicolo Tartaglia (1506–57), who accused Cardan of plagiarizing the cubic formula from his own teaching. As it happened, the priority for discovering the cubic formula goes neither to Cardan nor to Tartaglia, but to another Italian, Scipione del Ferro (1465–1526). Nevertheless, the formula is known as Cardan's formula.

Cauchy, Augustin-Louis (1789–1857) French mathematician whose lifelong goal was to put analysis on firm, rigorous foundations. Cauchy began his career as an engineer but soon turned to pure mathematics. Having refused to take an oath of allegiance to the new government after the fall of King Charles X in 1830, he was expelled from France and spent the next eight years in Italy, becoming a professor at the university of Turin, but returned home in 1838 to become a professor at the prestigious École Polytechnique. In his influential book *Cours d'analyse (Course in analysis,* 1821) he defined the concepts of continuity and differentiability in a precise, rigorous manner (this also included the limit concept, which was still rather vaguely understood in Cauchy's time); the modern "ε-δ" definition of these concepts is due to him. He was also a founder of the theory of functions of a complex variable (functions in which both the independent and dependent variables are complex numbers). A pair of equations known as the Cauchy-Riemann equations sets the conditions under which such a function is differentiable.

Cavalieri, Bonaventura (1598–1647) Italian mathematician who was an early pioneer of the ideas that would later lead to the invention of calculus. He was a disciple of Galileo and was appointed professor of mathematics at Bologna university in 1629. Cavalieri developed a crude method, known as "the method of indivisibles," to find the area of various shapes. Regarding the shape—say a segment of a parabola—as made up of infinitely many thin lines (in reality strips of infinitely small width) and summing up their areas, he was able to find the total area by a series of complicated calculations. Although the idea was ingenious, its practical usefulness was limited because each shape required a different technique. It was only when Newton and Leibniz invented the integral calculus that such problems could be solved using general methods. Cavalieri also promoted the use of logarithms following their invention by Napier in 1614.

Clairaut, Alexis-Claude (1713–65) French mathematician and theoretical astronomer. At the age of 10 he already studied calculus, and at 18 he published his first book. He worked on differential equations and applied them to celestial mechanics (the equation $xy' - y = f(y')$, where f is a function of the derivative y', is named after him). Clairaut was a member of the French expedition to Lapland to determine the shape of the Earth (which was found to be flattened at the poles, as Newton had predicted). In astronomy he worked on the theory of the Moon's motion, and he calculated to within one month the date of Comet Halley's return in 1759.

De Morgan, Augustus (1806–71) English mathematician, cofounder (with George Boole but independently of him) of symbolic logic. He was born in India to English parents, but he got his education at Cambridge, becoming the first professor of mathematics at University College in London. His most influential book was *Formal Logic* (1847), in which he devised algebraic symbols to propositional statements such as "if p then q" (written today $p \Rightarrow q$), "not p" (~p), "p or q" ($P \cup q$), and so on. Even more importantly, he introduced operational rules that allow us to manipulate these symbols in an algebraic manner—the essence of symbolic logic. De Morgan also wrote numerous articles on the history of mathematics and one satirical work, *An Encyclopedia of Paradoxes,* in which he parodied many false concepts common in his time.

Descartes, René (1596–1650) French philosopher and mathematician who invented analytic geometry. Descartes was a mercurial figure who switched careers more than once. He began his professional life as a soldier, often changing his allegiance and fighting for whichever side needed him most (nowadays we would call him a mercenary). Then one night he had a vision that God entrusted him with the key to unlock the secrets of the universe. He turned to philosophy, in which he would exercise enormous influence on subsequent generations of thinkers. He believed in a rational world in which everything flows from a cause; his motto was, "I think, therefore I am." Though primarily a philosopher, he expounded many physical theories, few of which stood the test of time. His theory of vortices, according to which space is filled with a thin fluid whose constant whirlpool-like motion causes the planets to move, was quite popular in Europe and for a while competed with Newton's theory of gravitation. Descartes also wrote on optics and physiology, but it is his invention of coordinate or analytic geometry that made his name immortal in mathematics. The idea of representing a point in the plane by two numbers ("coordinates")—its distances from two fixed lines ("axes")—is said to have come to him while lying in bed late one morning and watching a fly move across the ceiling. By assigning every point a pair of numbers, and regarding these numbers as variables, Descartes could express a geometric curve as a relation between two variables; in other words, he could represent the curve by an equation and thus use algebraic methods to investigate its properties.

This process also worked in reverse: an algebraic equation could be interpreted as a curve in the plane. Thus Descartes, in essence, united algebra and geometry, the two major branches of mathematics which, up until then, were almost entirely separate disciplines. We should mention that his coordinate system was not always rectangular (that is, his axes were not necessarily perpendicular to each other); in fact, he used different axes according to the curve under investigation. And he used only one quadrant (that is, he allowed only positive values for the variable). His exposition of coordinate geometry appeared in a relatively small work, *La géométrie,* which was published as an appendix to his major work, *Discourse on the Method of Reasoning* (1637), but it forever changed the course of mathematics.

Dirichlet, Peter Gustav Lejeune (1805–59) German mathematician. At the young age of 23, Dirichlet was appointed professor of mathematics at the University of Berlin. In 1855 he succeeded Gauss at the university of Göttingen, the world-renowned center of mathematical research up until World War II.

Dirichlet's work covered a wide range of subjects from analytic number theory (the study of the properties of integers with the help of calculus) to partial differential equations and their application to physics. He proved a famous theorem named after him: any arithmetic progression a, a + d, a + 2d, . . ., where *a* and *d* are relative prime (that is, have no common divisor other than 1), contains infinitely many primes. In analysis he stated the conditions sufficient for a Fourier series to converge (see FOURIER, JEAN-BAPTISTE-JOSEPH). His work on this subject led him to a new definition of function, essentially the one we use today.

Escher, Maurits Cornelis (1898–1972) Dutch artist who became famous for his mathematically related prints. A visit to the Alhambra, the Moorish palace in Granada (Spain) in 1936 turned Escher into an admirer of the elaborate geometric designs of Islamic art. These designs are based on the idea of *tessellation* (tiling)—filling the entire plane with the repetitions of a single motif, without overlap or empty space in between. Usually the motif was an abstract geometric shape—typically an eight-pointed star—but Escher gave it a human touch by choosing reptiles, birds, or fish as his motif. Among his most famous prints is one showing black and white horses marching in opposite directions, the spaces between the white horses becoming black horses, and vice versa. Escher was fascinated with infinity, endless cycles, and symmetry in all its mathematical aspects. Although not formally trained in mathematics (he barely passed his high school final mathematics exam), he has exercised an enormous influence on contemporary mathematics. The ideas depicted in his prints gave rise to new areas of mathematical research, and several international conferences, attended by mathematicians, artists, and graphic designers are devoted to his legacy.

Euclid (ca. 300 B.C.E.) Although Euclid did not play a direct role in the history of calculus, we include him here because of his all-important role in the history of mathematics in general. We know almost nothing about his life, except that he lived in Alexandria, the intellectual center of the Hellenistic world, where he may have been chief librarian of the great library at the Alexandrian academy. His *Elements* (in its full title, *The*

Thirteen Books of Euclid's Elements) is a compilation of all that was known in mathematics at his time. Euclid organized this body of knowledge into an orderly system of *axioms* (basic statements whose truth seems to be self-evident, such as the statement that between two points one and only one line passes), *definitions,* and *propositions* (in modern language, *theorems*)—statements whose truth must be proved based on the axioms and earlier propositions whose truth had already been established. This mode of reasoning became the standard in mathematics for the next two thousand years and is still the basic manner in which we think of mathematics today. Most of the 465 propositions in the *Elements* deal with geometry, but there are also many theorems in number theory and algebra (expressed in geometric language). The *Elements* had an enormous influence on subsequent generations of scientists; many great mathematicians were turned on to the subject after reading and mastering Euclid, among them young Newton.

Eudoxus of Cnidus (ca. 408–355 B.C.E.) Greek scientist who formulated the principle of exhaustion:

> *If from any magnitude there be subtracted a part not less than its half, from the remainder another part not less than its half, and so on, there will at length remain a magnitude less that any preassigned magnitude of the same kind.*

Simply put, this is a statement of the infinite divisibility of any magnitude, and it is the first seed from which the calculus would evolve more than two thousand years later (Archimedes would use this principle to compute the area of a parabolic segment, and to devise a method for approximating the value of π to any desired accuracy). Eudoxus also wrote on the theory of proportions—an important subject in early Greek mathematics—and on geography and astronomy. He was perhaps the first to propose that the heavenly bodies are carried around the Earth in spheres, a system that would be greatly elaborated upon by subsequent generations of Greek astronomers.

Euler, Leonhard (1707–83) Swiss mathematician, one of the most prolific of all time. There was hardly a branch of mathematics— pure or applied—in which Euler did not leave his mark, and his name is associated with even more formulas than Newton.

His combined writing—not yet published in full—is estimated to fill 70 volumes. Euler made significant discoveries in number theory, analysis, geometry, and topology, as well as in several branches of physics, notably mechanics and hydrodynamics. He was particularly adept in taking an equation and manipulating it in ways that produced numerous new results. His methods were not always rigorous, and occasionally he blundered; but more often he produced some of the most breathtaking formulas in mathematics, such as the equation $e^{i\pi} + 1 = 0$, which connects the four most important constants of mathematics (0, 1, e, and π) with the three most important operations (addition, multiplication, and exponentiation). Another of his achievements was to find the sum of the infinite series $1/1^2 + 1/2^2 + 1/3^2 + \ldots$, which had eluded some of the greatest mathematicians before him. Using an entirely intuitive approach, he found the sum to be $\pi^2/6$. In fact, he managed to find the sums of $\sum\limits_{n=1}^{\infty} 1/n^k$ for all even values of k up to 26, an amazing feat (the sum for odd powers is still not known). He also discovered the famous formula $V - E + F = 2$ connecting the number of vertices, number of edges, and number of faces of a simple polyhedron (a solid with no holes)—one of the first results in the branch of mathematics known as topology. In addition, he investigated numerous differential equations and devised methods for solving them. His great work *Introductio in analysin infinitorum* (Introduction to the analysis of the infinite, 1748) is regarded as the foundation of modern mathematical analysis. He also pioneered the application of the function concept to complex variables, and he devised many of the mathematical symbols that we use today, among them π, e, i, Σ, and $f(x)$.

Fermat, Pierre de (1601–65) French mathematician. Fermat was a lawyer by profession and pursued mathematics merely as a pastime, but his achievements were outstanding. He is regarded as the founder of modern number theory who discovered and proved many of its fundamental theorems. His most famous statement, known as Fermat's Last Theorem, says that it is impossible to find three integers x, y, and z that satisfy the equation $x^n + y^n = z^n$ for any integer n greater than 2 (we exclude the trivial case when $x = y = z = 0$). Fermat scribbled this statement (in words rather than as an equation) in the margin of a book he owned, adding, "I have found a wonderful proof of this, but unfortunately the margins are too small to give it." This enigmatic statement soon became the most celebrated unsolved problem in mathematics; numerous mathematicians—professionals and amateurs, and cranks as well—have attempted to prove it without success. The theorem was finally proved in 1995 by Andrew Wiles of Princeton University; the proof took 130 printed pages. Fermat also formulated an important principle in optics, according to which a ray of light always follows the path of shortest time (rather than shortest distance). In analysis, he devised a method for finding the maximum and minimum values of a function, and he found the antiderivative of the family of functions $1/x^n$ for all integral values of n except -1 (this last case leads to the logarithmic function)—all this several decades before the invention of calculus by Newton and Leibniz.

Fibonacci, Leonardo (**Leonardo of Pisa,** ca. 1170–1250) Italian mathematician. His book, *Liber abaci* ("the book of the calculator"), published in Pisa in 1202, was influential in spreading the Hindu-Arabic numerals in Europe. In this work he used algebra to solve a variety of equations and problems from geometry. His name is mostly remembered for the sequence 1, 1, 2, 3, 5, 8, 13, 21, . . ., in which every number beginning with the third is the sum of the two numbers preceding it. Fibonacci arrived at this sequence in connection with the way a pair of rabbits produces offspring, but the series turns up in various other situations, among them the arrangement of seeds in a sunflower and the growth pattern of a pine cone. The series also enjoys many unusual mathematical properties and is the subject of ongoing research.

Fourier, Jean-Baptiste-Joseph, baron de (1768–1830) French mathematician. In the tradition of many of France's great scientists, Fourier at first wished to pursue a military career, but he was rejected because of his lowly social status (his father was a tailor). Nevertheless, throughout his lifetime he continuously served his country as a civil servant. He joined Emperor Napoleon Bonaparte in his military campaign in Egypt, where he was put in charge of the French army workshops. Back in France, he became prefect of the district of Grenoble, in which role he supervised the government road construction and drainage operations. And if that were not enough, he was also appointed secretary of the prestigious Institut d'Egypte, and in 1809 he completed a major book on ancient Egypt. While fulfilling all these duties, he still found the time to do major research in algebra and analysis. His most famous result, known as *Fourier's Theorem,* states that every "reasonably behaving" periodic function can be expressed as an infinite series of sine and cosine terms, whose periods are integral divisors of the period of the function under consideration. For example, the function $f(x) = x$, regarded as a periodic function over the interval $-\pi < x < \pi$, can be written as a series of sine terms: $f(x) = 2[(\sin x)/1 - (\sin 2x)/2 + (\sin 3x)/3 - + \ldots]$. Fourier discovered this theorem in connection with his work on the heat conduction in solids and used it to solve the partial differential equation governing the process. *Fourier series* (also called trigonometric series) have found numerous applications in every branch of physics, from acoustics and earthquake analysis to wireless communication and quantum mechanics.

Galois, Evariste (1811–32) French mathematician and founder of modern group theory. His chief work was in finding the conditions under which an algebraic equation may be solved. Like his contemporary, the Norwegian NIELS HENRIK ABEL, Galois had difficulties getting the recognition he deserved for his pioneering work, partly because he was a political activist at a time (after the French Revolution) when any such activity could put a person's life in danger. He was killed in a duel over a trifling matter at the age of 21. Sensing the inevitable, he spent his last night summing up his many contributions to group theory.

Gauss, Carl Friedrich (1777–1855) German mathematician and physicist, regarded as one of the greatest scientists of all time. At the young age of 17, Gauss already had made a major discovery: that a 17-sided regular polygon can be constructed using only straightedge and compass; since the Greek era, it had been assumed that this could be done only with regular polygons of 3, 4, 5, and 15 sides and polygons obtained from these by repeatedly doubling the number of sides. This discovery so much impressed young Gauss that he made up his mind to pursue mathematics as a profession (earlier he had considered becoming a linguist). Like Euler before him, there was hardly a branch of mathematics in which Gauss did not leave his mark, from number theory and geometry to the calculus of several variables and of complex variables. He was the first to recognize that Euclid's Fifth Postulate (the "parallel axiom") could be replaced by alternative statements, thus giving rise to non-Euclidean geometry. Gauss also developed a branch of mathematics known as differential geometry—the application of calculus to the study of curved surfaces. He gave the first full and correct proof of the Fundamental Theorem of Algebra, which says that every polynomial with complex coefficients (which includes polynomials with real coefficients as a special case) has at least one root, or zero, in the system of complex numbers. In addition, Gauss did major work in physics, particularly in electromagnetism (the unit of the magnetic field is named after him). And if all this were not enough, he also left his mark on astronomy, devising a new method of computing the orbits of celestial bodies from only three observations. A statue of Gauss, standing on a 17-sided pedestal to honor his first major discovery, stands in his native town of Brunswick.

Germain, Sophie (1776–1831) French mathematician, one of only a handful of women mathematicians up until the 20th century. She was largely self-educated until she was 18, but even then could not enter a university: the newly founded École Polytechnique in Paris did not admit women. She had to confine her studies to lecture notes provided by others. Nevertheless, her interest in mathematics grew to the point where she submitted a thesis to the institute under the disguise of a male name. Her work greatly impressed Lagrange, who became her private tutor. Germain's main interests were in number theory, in which she worked on Fermat's Last Theorem (the famous conjecture that the equation $x^n + y^n = z^n$ has integer solutions $x, y,$ and z only for n = 1 and 2; it was only proved in 1995). Her contribution to this problem was largely ignored. Germain was also interested in the mathematical theory of elasticity, in which she investigated the vibration patterns of elastic plates. Despite her talents, she never received an academic degree.

Green, George (1793–1841) English mathematician and physicist. Green was largely self-taught. In a paper published in 1828 he introduced what would be known as *Green's Theorem,* a relation between a triple integral over the volume of a solid and a double integral taken over its boundary surface (this subject is studied in multivariable calculus). In this work he also introduced the *Green function,* a "higher" function that has found wide applications in modern applied mathematics. His 1828 paper remained largely unknown until it was resurrected by the famous physicist William Thomson (better known as Lord Kelvin) after Green's death.

Gregory, James (1638–75) Scottish mathematician and one of the early pioneers of calculus. Gregory studied infinite series and was the first to use the terms *convergence* and *divergence.* He discovered a number of new series, notably the series for the arctangent function, $\tan^{-1} x = x - x^3/3 + x^5/5 - x^7/7 + - \ldots$. Putting $x = 1$ in this series, he obtained the famous formula $\pi/4 = 1 - 1/3 + 1/5 - 1/7 + - \ldots$, still regarded as one of the most beautiful in mathematics (Leibniz rediscovered this series a few years later, and it is sometimes known as the Gregory-Leibniz series). Gregory was also the first to make a distinction between *algebraic* and *transcendental functions.* In astronomy, he proposed to use the occasional passage of the planets Mercury and Venus in front of the Sun to calculate the Astronomical Unit (the Earth–Sun distance), an idea that was later perfected by Halley and first implemented during the 1761 transit of Venus. He also proposed that a curved mirror could be used instead of a lens to build a telescope; it fell to Newton to actually build the first reflecting telescope.

Hamilton, William Rowan (1805–65) Irish algebraist who was the first to realize that mathematical operations need not necessarily follow the basic rules of arithmetic, which up until then were also the basic rules of algebra (after all, algebra was regarded as an extension of arithmetic to variable quantities, symbolized by letters but still representing numbers). In 1833 Hamilton came up with a representation of complex numbers that would once and for all rid these numbers of the air of mystery that was still surrounding them. He replaced the complex number a + ib, where *a* and *b* are real numbers and i = $\sqrt{-1}$, by an *ordered pair* (a, b) subject to certain algebraic rules. For example, he defined addition of two such pairs by the equation (a, b) + (c, d) = (a + c, b + d), which is equivalent to the familiar law (a + ib) + (c + id) = (a + c) + i(b + d). A slightly more complicated rule defined multiplication. The crucial step came when he introduced the rule (0, 1) × (0, 1) = (–1, 0), which is simply the rule $i^2 = -1$ in disguise. His paper on the subject greatly impressed the Irish Academy of Science and was a landmark in modern algebra, for it showed that mathematical systems can be constructed using arbitrary rules, as long as these rules are internally consistent.

Hamilton then tried to extend the system of complex numbers to three dimensions. He succeeded in doing so as far as addition was concerned, but failed with multiplication. For more than 15 years he struggled to find a way to multiply three-dimensional complex numbers, but to no avail. Then one day in 1843, while crossing a bridge over a canal in Dublin, an idea struck him: he had to give up the commutative law of multiplication. His basic units (later to be called vectors) were the quartet of symbols *1, i, j,* and *k.* The symbols *i, j,* and *k* followed the rules $i \times i = j \times j = k \times k = -1$. and thus behaved like the imaginary unit i = $\sqrt{-1}$. To these he added the rule $i \times j \times k = -1$. So far so good. But the breakthrough came when Hamilton realized that in order to get a consistent system, these symbols had to follow the *anticommutative* laws $i \times j = -j \times i = k$, $j \times k = -k \times j = i$, and $k \times i = -i \times k = j$. This was the birth of noncommutative algebra and, by extension, of abstract algebra—an algebra that can be applied to nonnumerical quantities. Hamilton then extended his theory to any linear combination of the symbols *1, i, j,* and *k,* that is, to quantities of the form a**1** + b**i** + c**j** + d**k,** where *a, b, c,* and *d* can be any real numbers. He called these quantities *quaternions;* later they would be called *vectors.* Hamilton also did significant work in theoretical mechanics and defined a quantity, called the *Hamiltonian,* which in essence replaced Newton's notion of force with energy as the fundamental quantity of a mechanical system.

In 1835 Hamilton was knighted, adding the title *Sir* to his name—one of a very few scientists to be thus honored

(Newton was another). A plaque on the Brougham Bridge in Dublin commemorates his discovery of noncommutative algebra. Sadly, his last years were spent in decline, due to an unhappy marriage and alcoholism.

Heaviside, Oliver (1850–1925) English physicist who mainly worked on the transmission of radio and telephone signals. He investigated the effects of inductance, capacitance, and impedance on transmission lines and derived the equation that governs the transmission of telegraph signals. He later extended his work to wireless telegraphy. Heaviside developed many of the mathematical tools that are now in standard use in the field, among them the use of complex numbers to describe alternating currents. He found an ingenious method for solving various differential equations that arise in electricity; his method became known as *operational calculus.* For example, the second-order differential equation $y'' - y = 0$ can be written as $D^2 y - y = 0$, where D is the differential operator d/dx; D^2 stands for d^2/dx^2. Heaviside treated D as if it were an ordinary algebraic quantity, so he rewrote the equation as $(D^2 - 1)y = 0$. Dividing both sides by y, we get the "operational equation" $D^2 - 1 = 0$, or $D^2 = 1$, whose "solutions" are $D = 1$ and -1. Remultiplying each solution by y, we get the two first-order differential equations $y' = y$ and $y' = -y$, whose solutions are $y = Ae^x$ and $y = Be^{-x}$, respectively, where A and B are arbitrary constants. Therefore the general solution of the original equation is $y = Ae^x + Be^{-x}$. Heaviside was a master in manipulating the D symbol, but his method was frowned upon by professional mathematicians as lacking rigorous foundation. In his defense, he maintained that the end justifies the means—after all, his methods did work and enabled scientists to solve numerous differential equations with ease and efficiency. Heaviside was self-educated and never held an official academic position, which makes his achievements all the more remarkable. He died in neglect and poverty.

Hermite, Charles (1822–1901) French mathematician who proved that the number e, the base of natural logarithms, is *transcendental* (that is, it is not a solution of any polynomial equation with integer coefficients). This discovery (1873) led directly to Ferdinand Lindemann's more famous discovery that π too is transcendental (1882). In contrast to the early success stories of many famous mathematicians, Hermite's fame rose only slowly; in fact, he was a rather mediocre university student, yet his later contributions were significant. He worked on elliptic functions and applied them to finding the general solution of the quintic equation (a polynomial equation of degree 5). As other mathematicians of his era often did, he applied his analytic methods to number theory (the study of integers), even though these fields are at the extreme opposites of the mathematical spectrum. The nonglamorous beginning of his career was more than compensated for when in 1870 he became professor of mathematics at the Sorbonne, France's most prestigious university.

Hilbert, David (1862–1943) German mathematician, one of the towering figures of late 19th- and early 20th-century mathematics. At the young age of 33 he became chair of the mathematics department at the university of Göttingen, the world-renowned center of mathematical research up until World War II. Hilbert's research covered almost every area of mathematics then known, making him one of the last universalist scientists. He felt as much at home in mathematical physics as in number theory, two areas of research at the extreme opposites of the mathematical spectrum. He was also a master teacher who exercised a profound influence on his many students and colleagues.

It is almost impossible to do justice to Hilbert's enormous contributions in a few lines. His goal was to put all of mathematics on a firm, logical foundation, devoid of any trace of intuition. His formalistic view stirred up much debate and gave rise to three schools of mathematical thought: the formalist, the intuitionist, and the logistic schools. Hilbert never stayed too long in the same field of research: when he felt he had made a significant impact on one field, he abandoned it and moved on to another. In mathematical physics he developed the concept of an infinitely multidimensional space, now called *Hilbert space,* in which the familiar two-dimensional Pythagorean Theorem $c^2 = a^2 + b^2$ is replaced, first by an *n*-dimensional expression of the form $c^2 = a_1{}^2 + a_2{}^2 + \ldots + a_n{}^2$, and then by the infinite series $c^2 = a_1{}^2 + a_2{}^2 + \ldots + a_n{}^2 + \ldots$, provided the series converges. This abstract idea has proved of enormous value to the development of modern physics, in particular quantum theory.

At the Second International Congress of Mathematicians, held in Paris in 1900, Hilbert challenged the mathematical community with 23 problems whose solution he felt would have great implications for 20th-century mathematics. Some of Hilbert's problems remain unsolved; all of them have had a profound influence on mathematical research, opening up new areas of research and extending existing ones, which was precisely Hilbert's goal in presenting his problems.

Hipparchus of Nicaea (ca. 190–120 B.C.E.) Greek mathematician and astronomer who is regarded as the founder of trigonometry. As an astronomer, he was chiefly interested in *spherical trigonometry* (the study of triangles drawn on a spherical surface), but he also established most of the formulas of plane trigonometry that are now the core of any college course (all of which he derived geometrically). He also computed the first trigonometric table, essentially a table of sines. Around 130 B.C. Hipparchus compiled a catalog of 850 stars, being the first ever to do so. In his catalog he used a system of coordinates similar to geographical longitude and latitude to locate the position of any star on the celestial dome; these coordinates are now called *right ascension* and *declination,* respectively. Hipparchus also discovered the *precession of the equinoxes*—a steady shift in the direction of Earth's axis of rotation that takes about 25,800 years to complete. And he was the first to arrive at a reasonable value for the distance between the Earth and the Moon; he estimated it to be between 71 and 83 Earth radii (the actual value varies from 56 to 64). Unfortunately, as with most scientists of his time, very little is known about his life.

L'Hospital, Guillaume-François-Antoine, marquis de (also L'Hopital or L'Hôpital, 1661–1704) French mathematician who wrote the first textbook on calculus. He studied under Johann Bernoulli and was influential in disseminating the newly invented calculus throughout Europe. A famous theorem, known as L'HOSPITAL'S RULE, was actually discovered by Bernoulli, but by mutual consent it was

L'Hospital who first published it. The rule allows one to find the limit of an indeterminate expression of the form 0/0 or ∞/∞, as in the expression (sin x)/x as x $\to$ 0.

Kepler, Johannes (1571–1630) German astronomer and mathematician. Kepler was a mystic who believed that number and shape are the keys to understanding the universe. In this he was perhaps the last of the Pythagoreans, who had adopted these ideas two thousand years earlier. Nevertheless, by sheer perseverance, and after 30 years of search (which included ruminations into the theory of musical harmony), he finally discovered the laws governing the motion of the planets. The first of Kepler's three laws says that the planets move around the Sun in ellipses, the Sun being at one focus of each ellipse. Thus Kepler in effect brought to a close Greek astronomy, with its belief in circular orbits around a stationary Earth. Kepler was also an early pioneer of the method of indivisibles, a crude method of "slicing" a solid into infinitely many thin layers, each of infinitely small width, and summing up their volumes. This method eventually evolved into the integral calculus, but it had already been used half a century before Newton and Leibniz invented the calculus. Kepler used this method to find the volume of various wine casks in an attempt to force government taxers to gauge the wine content of barrels in a correct manner.

More significantly, his three laws of planetary motion put the conic sections (the ellipse, parabola, and hyperbola) back on the mathematical scene, after having been nearly forgotten since the end of the Greek era.

Lagrange, Joseph-Louis, comte de (1736–1813) Italian-born French mathematician. Lagrange was born in Turin and became professor of mathematics at the Royal Artillery School there when only 19. In 1766 he moved to Berlin, where he succeeded Euler as director of the mathematical section of the Berlin Academy of Sciences. In 1787, upon the invitation of Louis XVI, he settled in Paris and became a member of the French Academy of Sciences. In this capacity he chaired the commission that introduced the metric system to France (and later to most of the world) following the French Revolution. In his mathematical work, Lagrange in effect made analytical mechanics a branch of applied mathematics. He reformulated Newton's three laws of motion into a set of differential equations, in which the energy of a system, rather than the forces acting on it, played the central role. He used this new interpretation in his chief work, *Méchanique analytique* (1788), a treatise on analytical mechanics that became a model of rigorous exposition to future scientists. Lagrange started this work when he was 19 and finished it at 52; the entire work does not contain a single illustration—Lagrange let the equations speak for themselves. He also did major work in number theory and in continuum mechanics; in the latter field he formulated a partial differential equation that governs the motion of fluids, known as Lagrange's equation. The notation f′ for the derivative was due to him (he denoted a function by fx and its derivative by $f'x$—the precursors of our f(x) and f′ (x)).

Lambert, Johann Heinrich (1728–77) Swiss scientist who wrote on a wide range of subjects, including mathematics, astronomy, and optics (he was the first to propose a precise method for measuring the intensity of light). In mathematics his main claim to fame was to prove that π is irrational; this in effect put an end to the age-old search for two integers, however large, whose ratio would be exactly equal to π. Lambert is often credited with introducing hyperbolic functions into mathematics, but this honor goes to an Italian Jesuit, Vincenzo Riccati (1707–75), who introduced them in 1757.

Laplace, Pierre-Simon de Laplace, marquis de (1749–1827) French mathematician and theoretical astronomer. As with his countryman Lagrange, Laplace was courted by Emperor Napoleon Bonaparte. He became member of the Senate and later its vice president. In this he followed a long line of distinguished French scientists who also served their country in public and military service. His major work, the five-volume *Méchanique céleste* (Celestial mechanics) was hailed as the greatest work on the subject since Newton's *Principia*. Laplace studied the stability of the solar system based on Newton's universal law of gravitation. He proposed his famous nebular hypothesis, according to which the solar system was formed from a vast swirling gas cloud that gradually condensed under its own gravity, forming the Sun and planets; essentially, this is still the view held today. He introduced into physics the notion of potential and formulated a partial differential equation, named after him, that described the gravitational or electric field under prescribed conditions. Laplace also did pioneering work in probability; the Laplace distribution is named after him.

Legendre, Adrien-Marie (1752–1833) French mathematician. His main work was in number theory, celestial mechanics, and the study of a class of higher (nonelementary) functions known as elliptic functions. He introduced the Legendre polynomials, a family of solutions to a second-order differential equation that frequently shows up in physics. In 1794 he wrote a popular textbook on geometry, in which he gave a proof that π is irrational (though he was not the first to prove it; he was, however, the first to prove that π^2 is irrational). Legendre also gave the first satisfactory treatment of the method of least squares, which had originally been discovered by Gauss. As with many of his French colleagues, Legendre served in various public functions, among them as head of the governmental department for standardizing weights and measurements, founded in 1794 during the introduction of the metric system. Later, in response to his refusal to comply with the government's attempts to dictate its policies to the French Academy of Sciences, he was deprived of his pension and spent his last years in poverty.

Leibniz, Gottfried Wilhelm, Freiherr von (1646–1716), German philosopher and mathematician and Newton's great rival in the invention of calculus. Leibniz began his career as a diplomat in the service of his patron, the elector of Mainz; his mission was to restore European peace after the Thirty Years' War. In this capacity he traveled to France and England and made many acquaintances, some of whom would later play a role in his priority dispute with Newton. He was also involved in a failed attempt to reunite the Catholic and Protestant churches. As a philosopher, Leibniz advanced the idea that everything in the world—physical as well as spiritual—follows from the interaction of innumerable tiny units he called *monads*. He attempted to develop a formal system of logic in which all deductions could be made in an algorithmic, computational manner (this idea would be taken up a century later by the English mathematician George Boole, the founder of symbolic logic). Leibniz was also the first to recognize the significance of the binary base—the number system that uses only two numerals, 0 and 1, the basis of modern computers; his interest in this base, however, was more philosophical than practical, seeing it as a gift from God to humanity. But by far his greatest contribution to mathematics was his invention of the calculus, which he developed during the decade 1666–75 independently of Newton. Whereas the reclusive Newton withheld publication of his own results, Leibniz published his invention in 1684, precipitating one of the most ugly priority disputes in the history of science. The two great minds had arrived at the same results—in particular, they both discovered what is known today as the *Fundamental Theorem of Calculus* (the inverse relation between differentiation and integration), but their approach and notation were different, with Leibniz proposing the more efficient "d" notation (see DERIVATIVE). Today we give Newton and Leibniz equal credit for the invention of calculus, an achievement that forever changed the course of mathematics.

Maclaurin, Colin (1698–1746) Scottish mathematician who was influential in disseminating the newly invented calculus throughout England. In his *Treatise of Fluxions* (1742) he attempted to give Newton's differential calculus ("fluxion" was Newton's word for derivative) a geometric foundation. This ran contrary to the trend that began to form in continental Europe, where mathematicians were trying to put the calculus on firm, logical foundations, but it made the subject a lot more accessible to English scientists. Ironically, Maclaurin's name is known today mainly for an infinite series that was actually discovered by his contemporary Brook Taylor (see MACLAURIN SERIES).

Napier, John (1550–1617) Scottish mathematician and the inventor of logarithms. His early life did not hint at any future mathematical greatness. He was a practical man who invented a variety of mechanical devices to improve the crop on the farm on which he lived; these included a hydraulic screw to control the level of water. He also showed an interest in military hardware and drew plans for building a huge artillery piece and even a submarine. If that was not enough, Napier was also a religious activist who got himself embroiled in many controversies. It is not known what led him to the idea of logarithms, on which he worked for 20 years. His tables, published in 1614, were received with great enthusiasm by the scientific community, for they greatly reduced the labor of numerical computing (logarithms allow us to replace multiplication and division by addition and subtraction). In his original logarithms he did not use a base in the modern sense, but with the help of the English mathematician Henry Briggs, who traveled to Scotland to meet Napier, they reworked the tables and made them into base 10 ("common") logarithms. In this form they remained virtually unchanged, until the advent of the electronic handheld calculator in the 1970s made them obsolete. Napier also invented the Napier rods—a sort of mechanical calculator—and he devised a set of rules known as "Napier analogies" for use in spherical trigonometry. And he advocated using the decimal point to separate the integral part of a number from its fractional part. But it is his invention of logarithms that made his name immortal.

Newton, Sir Isaac (1642–1727) English mathematician and physicist, by general consensus one of the three greatest scientists of all time (the others are Archimedes and Einstein). Newton's early life was beset by misfortunes. His father died shortly before Isaac was born; his mother soon remarried, only to lose her second husband too. Young Newton was thus left in the custody of his grandmother. In 1661 he entered Trinity College (part of Cambridge University), where his mathematical genius flourished. He studied many of the classic works on mathematics, including Euclid's *Elements* and Descartes's *La Géometrie*—none of which is easy reading even today. The fact that he studied these works on his own, with little help from the outside, set the stage for his future character—a reclusive man who was reluctant to share his thoughts with others. Indirectly it would contribute to his bitter priority dispute with Leibniz over the invention of the calculus.

In 1665 Cambridge University closed its doors due to the outbreak of the Great Plague. Newton returned to his family's farm, where he enjoyed two years of complete freedom to shape his scientific ideas; later he would refers to this period as his "prime years." Newton's first major discovery was the expansion of $(a + b)^n$ into powers of a and b when n is a negative integer or a fraction (the case when n is a positive integer had been known for a long time and involves the Pascal Triangle); the expansion in these cases is an *infinite series.* At about the same time Newton began to shape his thoughts on gravitation, reportedly triggered by seeing an apple fall from the tree (there is no evidence, however, that this actually happened). He also speculated on the nature of light and discovered the splitting of white light into its spectrum of rainbow colors. And if these discoveries were not enough, he also worked out his "method of fluxions"—his differential and integral calculus. Unlike his rival Leibniz, Newton was always guided by physical intuition; he thought of a function as a relation between two variables, each of which "flows" continuously with time (hence the word *fluxion*). But his reluctance to publish his discoveries, while Leibniz published his own, precipitated a bitter priority dispute between the two, and the aftershocks lasted well after both men were dead. Newton's work on gravitation was published in his great work, *Philosophiae naturalis principia mathematica* (Mathematical principles of natural philosophy, 1687). The *Principia,* as it is commonly known, has had an enormous influence on subsequent generations of scientists and was hailed as the greatest work in science since Euclid wrote his *Elements* around 300 B.C.E.; it marked the beginning of the modern era in science. As for the calculus, a summary of it was not published until 1704 as an appendix to Newton's other great work, *Opticks,* but a full account had to wait until 1736, nine years after his death. Newton died at the age of 85 and was given a state funeral; he was buried at Westminster Abbey in London, where an ornate tombstone marks the site.

Pascal, Blaise (1623–62) French mathematician, physicist, and philosopher who was educated by his father, himself a mathematician. But the father, being a pedant, insisted that his son should first become acquainted with classical languages, so he forbade young Pascal to read any mathematics books.

Secretly, however, the 11-year-old Pascal read Euclid's *Elements,* which he mastered all by himself. At the age of 16 he wrote an original paper on conic sections that amazed Descartes. Pascal pioneered (with Fermat) the theory of probability, made major discoveries in geometry, and built a calculating machine that could add and subtract numbers of up to eight digits. The famous Pascal Triangle (a triangular array of numbers whose nth row gives the coefficients in the expansion of $(a + b)^n$ in powers of a and b) was not his invention, however; it had been known long before him. At 23 he turned to physics and discovered the law of hydrostatic pressure that bears his name. Then at 25 he suddenly lost interest in mathematics and science and spent his remaining years in religious penance. However, in one last flash of creativity, he found the area under the *cycloid,* thus anticipating the soon-to-be-discovered integral calculus.

Pythagoras of Samos (ca. 585–500 B.C.E.) We know almost nothing about the life of this legendary figure. The little we do know comes from scholars who lived hundreds of years after him, so much of it may be more legend than fact. Pythagoras founded a school of philosophy whose members were bound by a vow of secrecy, which is why almost nothing of his teachings survived in the original. The Pythagorean motto was "Number rules the universe," and by "number" was meant integers and ratios of integers. This belief was probably motivated by Pythagoras's discovery that the rules of musical harmony depend on the ratios of simple integers, such as 2:1 for the octave and 3:2 for the fifth (so called because it leads to the fifth note in the musical scale). So when one of the Pythagoreans—we do not know his name—found that the square root of 2 cannot be written as a ratio of two integers (today we call such a number an irrational number), the discovery came as an utter shock. Legend has it that the sect members, fearing that the discoverer might reveal this to the outside world, threw him overboard the boat they were sailing. The Pythagoreans were also the first to deal with number theory, to which they were led by their belief in numerical mysticism. As for the theorem that made Pythagoras's name immortal, he did not actually discover it, for we know from clay tablets that the Babylonians knew the theorem at least a thousand years before he did. But it was Pythagoras who first *proved* the theorem, thus extending its validity to all right triangles, not just some particular ones.

The Pythagorean school exercised an enormous influence on subsequent generations of scientists, an influence that lasted well over 2,000 years. When the astronomer Johannes Kepler, in the early 17th century, sought to discover the laws of planetary motion, he was at first guided by the Pythagorean ideas of musical harmony. It was only after 30 years of hard labor that he realized the fallacy of such an approach. Much of the Pythagorean number mysticism survives even today.

Riemann, Georg Friedrich Bernhard (1826–66) German mathematician who did groundbreaking work in non-Euclidean geometry, differential geometry, and the theory of functions of a complex variable. Riemann studied under Gauss and spent most of his short life at Göttingen, the world-renowned center of mathematical research up until World War II. In 1854 he gave a lecture entitled "Concerning the Hypotheses which Underline Geometry," in which he introduced the notion of an *n*-dimensional curved space; this concept would later become an essential element in Einstein's general theory of relativity. Riemann studied the infinite series $1 + 1/2^s + 1/3^s + \ldots$, in which the exponent *s* is a complex number u + iv. Riemann regarded this series as a function of *s*, known as the *zeta function* and denoted by $\zeta(s)$; he conjectured that all non-real zeros of $\zeta(s)$ have their real part equal to 1/2 (that is, they lie on the vertical line u = 1/2). To this day, Riemann's hypothesis remains unproved and is regarded as one of the greatest unsolved problems in mathematics (surprisingly, the zeros of this function are related to the distribution of prime numbers). To students of calculus, Riemann's name is familiar from the *Riemann sum,* which plays a key role in the definition of the definite integral.

Rolle, Michel (1652–1719) French mathematician who, like many of his colleagues, served his country in a military role. He wrote on algebra and geometry and gained a reputation as a critic of the newly invented calculus, which he claimed was "a collection of ingenious fallacies." He is remembered today mainly for the theorem that bears his name (see ROLLE'S

THEOREM), which was used by Lagrange to prove the more general *Mean Value Theorem*.

Simpson, Thomas (1710–61) English mathematician who is remembered chiefly for the rule named after him for approximating the area under a curve (see SIMPSON'S RULE). Simpson made a living as a weaver and taught mathematics as a side job. His life was marked by strange events: when 20, he married a 40-year-old widow who lived to be 102—exactly twice Simpson's life span. For a while he made himself a name as an astrologer, but after scaring a young girl with his forecast, the Simpsons were forced to flee. He wrote several mathematics books on a variety of topics, among them an early text on calculus (1737) and two books on probability. In 1745 he was elected to the Royal Society.

Taylor, Brook (1685–1731) English mathematician who is remembered mainly for his theorem, published in 1715, on the expansion of a function in an infinite power series (TAYLOR SERIES; TAYLOR THEOREM). Taylor also worked on the theory of vibrating strings, a subject that preoccupied many of the leading mathematicians of the 18th century, and he published two works on perspective.

Viète, François (1540–1603) French mathematician who was the first to introduce into algebra the use of letters to indicate constants and variables. As with many French scientists, notably Fermat and Descartes, he practiced mathematics not as a profession but as a recreation. And again like many of his French colleagues, he served his country in military duty, in this case as a cryptologist who deciphered many of Spain's communications during its war with France. Viète's most influential work was his *In artem analyticem isagoge* (Introduction to the analytic art, 1591), considered the earliest work on symbolic algebra (before him, algebraic operations were expressed verbally). He denoted known quantities by consonants and unknowns by vowels (our present custom of using *a, b, c,* etc., for constants and *x, y, z* for variables was introduced by Descartes in 1637). Viète also gave the basic rules of solving equations, such as moving a term from one side of the equation to the other, dividing an equation by a common factor, and so on. His introduction of symbolic algebra is considered one of the most important developments in the history of mathematics. Viète also made substantial contributions to trigonometry, in effect uniting it with algebra. He is chiefly remembered for a remarkable formula he discovered in 1593: $2/\pi = (\sqrt{2}/2) \cdot (\sqrt{2 + \sqrt{2}})/2 \cdot (\sqrt{2 + \sqrt{2 + \sqrt{2}}})/2 \cdot \ldots$, which expresses π as an infinite product of square roots of 2. This marks the first time that an infinite process was explicitly written as a formula; up until then, mathematicians preferred to avoid invoking the concept of infinity directly—a leftover from the Greek era. In his last years Viète got himself embroiled in an ugly controversy over the reformation of the calendar called for by Pope Gregory XIII. Viète bitterly opposed the reform, and in the process he made many enemies. He was also opposed to Copernicus's new heliocentric system. We see here the inner conflicts of a man who lived in a time of transition between the old world and the new.

Wallis, John (1616–1703) English mathematician and clergyman and a founder of the Royal Society. In 1649 he was appointed professor of geometry at Oxford University. In 1660 Charles II made him his royal chaplain (it was quite common for scientists in Wallis's time to have clerical assignments). Wallis did important work in algebra and geometry and was one of the pioneers of the differential and integral calculus just before Newton and Leibniz. His most notable book was *Arithmetica infinitorum* (Arithmetic of the infinite, 1655), which young Newton studied while a student at Cambridge. This work contains a remarkable formula, known as Wallis's product, for which he is chiefly remembered: $\pi/2 = (2/1) \cdot (2/3) \cdot (4/3) \cdot (4/5) \cdot (6/5) \cdot (6/7) \cdot \ldots$ Wallis represents the era of transition from the old, Greek mathematics, which was essentially geometric, to the new mathematics of algebra and analysis. He was the first to treat the conic sections as quadratic equations, rather than geometric objects. He also wrote on mechanics (1669) and on algebra (1685). He introduced the symbol ∞ for infinity and advocated the use of negative and fractional exponents to express ratios and radicals. Wallis was also a cryptologist who deciphered secret messages during the English Civil War, he pioneered teaching speech to deaf people, and he was one of the first professional mathematicians to write on the history of mathematics.

Weierstrass, Karl Theodor Wilhelm (1815–97) German mathematician whose work on functions set a new standard of rigor in mathematics. Weierstrass started his mathematical career late in life, having first studied law and finance and making a living as a mathematics teacher. He was almost 50 when he became a professor at the University of Berlin, but his subsequent work was outstanding. In 1861 he demonstrated that there exist functions which are everywhere continuous on an interval but nowhere differentiable. This discovery came as a shock to mathematicians, because it ran against "common sense" intuition. It demonstrated the need to base analysis, and mathematics in general, on strict logical foundations, devoid of geometric or other intuitive considerations; in particular, Weierstrass gave rigorous definitions of such fundamental concepts as function, continuity, limit, convergence, and differentiability. He based much of his work on infinite series in the complex plane and used them to define what is known as analytic functions (differentiable functions of a complex variable). He also worked on "higher" (nonelementary) functions known as Abelian integrals. His uncompromisingly high standards of rigor made him the "mathematical conscience" of his time and served as a model to future mathematicians.

SECTION THREE
CHRONOLOGY

ca. 1800–1600 B.C.E. ● Clay tablets from Mesopotamia show that the Babylonians had considerable knowledge of geometry and algebra and could perform complex numerical computations. They used a base-60 numeration system, and they knew the Pythagorean Theorem a thousand years before Pythagoras proved it.

ca. 1650 B.C.E. ● The Rhind Papyrus, a collection of 84 problems in arithmetic, geometry, and primitive algebra, is the earliest complete mathematics textbook to come to us. It was found in Egypt in 1858 and is now in the British Museum.

ca. 540 B.C.E. ● Pythagoras of Samos proves the Pythagorean Theorem. He also discovers that $\sqrt{2}$ is irrational.

ca. 360 B.C.E. ● Eudoxus formulates the method of exhaustion.

ca. 300 B.C.E. ● Euclid of Alexandria summarizes geometry and number theory in his *Elements,* the most influential book in the history of mathematics.

ca. 220 B.C.E. ● Appolonius of Perga writes on conic sections.

ca. 200 B.C.E. ● Archimedes of Syracuse uses the method of exhaustion to find the area of a segment of a parabola. He also discovers a procedure for approximating π to any desired accuracy; he uses it to show that π is between $3^{10}/_{71}$ and $3^{10}/_{70}$.

ca. 150 B.C.E. ● Hipparchus of Nicaea develops trigonometry and astronomy.

ca. 150 C.E. ● Claudius Ptolemaeus of Alexandria (commonly known as Ptolemy) writes his *Almagest,* the most influential work in mathematical astronomy until the 16th century. In it he gives the first systematic treatment of trigonometry, including a table of chords (essentially a table of sines).

ca. 250 ● Diophantus of Alexandria writes his *Arithmetica,* the first known book on algebra.

ca. 415 ● Death of Hypatia of Alexandria, the first woman mathematician.

529 ● Closure of the academies of Athens; the end of the Greek era of eminence in mathematics, philosophy, art, and literature.

ca. 600 ● The Hindus invent the base 10 numeration system, later to be called the Hindu-Arabic system.

ca. 830 ● Al-Khowarizmi writes his *Hisab al jabr w'al-muqua-balah* (Science of transposition and cancellation), an influential work on algebra (the modern word *algebra* evolved from the "al jabr" in the title). He also introduces the Hindu base 10 numeration system to Europe. The modern word *algorithm* is a corruption of his name.

1202 ● Leonardo Fibonacci ("Leonardo of Pisa") writes his *Liber abaci* (The book of the calculator), which popularized the Hindu-Arabic numeration system in Europe. He also discovers the series 1, 1, 2, 3, 5, 8, 13, 21, … named after him.

1482 ● First Latin printing of Euclid's *Elements*.

1527 ● Petrus Apianus shows the Pascal Triangle on the title page of his book *Rechnung* (Calculation).

1533 ● The first comprehensive modern treatise on trigonometry, *De triangulis omnimodis* (On triangles of every kind) by Regiomontanus, is published in Nürnberg.

1543 ● Nicolaus Copernicus's *De revolutionibus* (On the Revolutions) is published and brought to the author just hours before his death. In it Copernicus sets forth his heliocentric (Sun-centered) system, the most profound change in our view of the universe to date.

1544 ● Publication of Michael Stifel's *Arithmetica integra,* a treatise on numbers and algebra.

1545 ● Girolamo Cardan publishes his *Ars magna,* a treatise on algebra in which the author claims as his own the solution to the cubic equation, actually discovered earlier by Nicolo Tartaglia. A bitter priority dispute follows.

1569 ● Gerhard Mercator publishes his world map, based on a new projection he invented, that forever changes marine navigation.

1591 ● François Viète publishes his *In artem analyticem isagoge* (Introduction to the analytical art), the first work on symbolic algebra, in which he introduces letters to denote constants and variables.

1609 ● Johannes Kepler publishes his *Astronomia nova* (New astronomy), in which he announces the first two of his three laws of planetary motion; the third law was announced in 1619. This marks the beginning of modern mathematical astronomy.

1614 ● John Napier publishes his invention of logarithms, the single most important aid to computing until the advent of electronic computers.

1637 ● René Descartes introduces analytic (coordinate) geometry into mathematics, thereby uniting algebra with geometry.

1642 ● Death of Galileo Galilei and birth of Isaac Newton.

1654 ● Blaise Pascal and Pierre de Fermat develop probability theory.

1666–1676 ● Isaac Newton in England and Gottfried Wilhelm Leibniz in Germany independently invent the differential and integral calculus. Newton's invention is not published until 1704.

1684 ● Leibniz publishes his version of the calculus, precipitating a bitter priority dispute with Newton.

1687 ● Newton publishes his *Principia,* the single most influential book on physics in the history of science. In it he announces the Universal Law of Gravitation and formulates the three laws of motion named after him.

1696 ● The Bernoulli brothers solve the *brachistochrone* problem— to find the curve along which a particle slides down under the force of gravity in the least possible time; the curve is an inverted cycloid.

● L'Hospital publishes his *Analyse des infiniment petits* (Analysis of the infinitely small), the first textbook on calculus.

1718 ● Jakob Bernoulli publishes his work on the calculus of variations, in which one seeks a *function* (rather than a number) that maximizes or minimizes a definite integral.

1727 ● First use of the letter *e* for the base of natural logarithms (Euler).

● Sir Isaac Newton dies.

1748 ● Leonhard Euler publishes his *Introductio in analysin infinitorum,* considered the foundation of modern analysis. It contains numerous new results on infinite series, among them the formula $\sum_{n=1}^{\infty} 1/n^2 = \pi^2/6$.

1757 ● Vincenzo Riccati introduces hyperbolic functions.

1788 ● Joseph-Louis Lagrange publishes his *Mécanique analytique,* thereby making analytical mechanics a branch of applied mathematics.

1797 ● Lagrange publishes his *Théorie des fonctions analytiques,* in which the notations f and f' for a function and its derivative appear for the first time systematically.

● Caspar Wessel shows how complex numbers can be represented graphically as vectors.

1799 ● Carl Friedrich Gauss gives the first satisfactory proof of the Fundamental Theorem of Algebra.

1801 ● Gauss publishes his *Disquisitiones arithmeticae,* his major work on number theory.

1806 ● Jean Robert Argand shows how multiplication of complex numbers can be represented graphically as vector rotation.

1821 ● Augustin-Louis Cauchy defines the concepts of limit and continuity in terms of "getting arbitrarily close"; this comes close to the modern "ε-δ" definitions.

1822 ● Jean-Baptiste-Joseph Fourier publishes his *Théorie analytique de la chaleur* (Analytic theory of heat), in which Fourier series are introduced.

● Charles Babbage begins construction of his Difference Machine, considered the first mechanical computer.

1827 ● Cauchy publishes his *Calculus of Residues,* in which he extends the concept of definite integral to the complex plane, thus setting the foundation for the modern theory of functions of a complex variable.

● Gauss introduces differential geometry, the application of calculus to the study of curved surfaces.

1829 ● Nicolai Ivanovitch Lobachevsky and Janos Bolyai independently found non-Euclidean geometry.

1843 ● William Rowan Hamilton introduces quaternions (noncommutative extensions of complex numbers) into algebra. This marks the birth of modern abstract algebra.

1844 ● Joseph Liouville discovers the first transcendental number.

1847 ● George Boole publishes his *Mathematical Analysis of Logic,* marking the beginning of symbolic logic (also called Boolean algebra).

1854 ● Georg Friedrich Bernhard Riemann's lecture "Concerning the Hypotheses which Underline Geometry" introduces the concept of an *n*-dimensional curved space, later to play a key role in Einstein's general theory of relativity.

1858 ● Arthur Cayley introduces matrices into algebra.

1872 ● Felix Klein, in his Erlanger Program, puts forth a new view of geometry as a set of transformation groups.

1873 ● Charles Hermite proves that *e* (the base of natural logarithms) is transcendental.

1874 ● Georg Cantor creates modern set theory and uses it to define different classes of infinity represented by transfinite cardinals.

1882 ● Carl Louis Ferdinand Lindemann proves that π is transcendental, settling the age-old problem of constructing, by straightedge and compass, a square equal in area to a given circle. The proof shows that the construction is impossible.

1888 ● The American Mathematical Society is founded.

1896 ● Jacques-Salomon Hadamard and Charles de la Vallée Poussin independently prove the Prime Number Theorem, first conjectured by Gauss in 1792.

1900 ● David Hilbert poses his celebrated 23 problems at the Second International Conference of Mathematicians in Paris.

1910–1913 ● Bertrand Russell and Alfred North Whitehead publish their three-volume *Principia mathematica,* a grand attempt to base all of mathematics on a small set of axioms and logical principles.

1916 ● Albert Einstein publishes his general theory of relativity, in which Riemann's curved geometry plays a central role. It would fundamentally change our perception of space, time, mass, and gravity.

1931 ● Kurt Gödel publishes his "On Formal Undecidable Theorems of the *Principia Mathematica*," in which he demonstrates the existence of undecidable problems— statements whose truth or falsity cannot be proved within a given formal logical system.

1937 ● Alan Turing describes his "universal computing machine" (since known as the Turing machine), a theoretical computer that could be programmed to perform specific tasks.

1944 ● John von Neumann and Oskar Morgenstern develop game theory.

1945 ● ENIAC, the first fully automatic electronic digital computer, becomes operational at the University of Pennsylvania.

1961 ● Edward Lorenz founds chaos theory.

1971 ● The first electronic pocket calculator appears on the market.

1974 ● The first programmable pocket calculator is introduced by Hewlett-Packard Company.

1975 ● Mitchell Feigenbaum discovers a new fundamental mathematical constant, approximately 4.6692, that plays a role in chaos theory.

1977 ● Benoit Mandelbrot introduces the term *fractal* into mathematics.

1978 ● Laura Nickel and Curt Noll, two 18-year-old students from Hayward, California, discover the largest prime to date, $2^{21,701} - 1$, a 6,533-digit number. It takes them 440 hours of computer time.

1980 ● In a worldwide effort, mathematicians complete the classification of all finite simple groups.

1995 ● Andrew Wiles of Princeton University proves Fermat's Last Theorem; his paper, entitled "Modular Elliptic Curves and Fermat's Last Theorem," is 130 pages long.

1997 ● The largest prime to date is discovered: $2^{2,976,221} - 1$, a 895,932-digit number that would fill a 450-page book if printed. It is almost immediately superseded by two even larger primes: $2^{3,021,377} - 1$ (discovered in 1998) and $2^{6,972,593} - 1$ (1999).

2001 ● Dwarfing the primes mentioned above, a team lead by Michael Cameron, Scott Kurowski, and George Woltman discovers the gargantuan prime $2^{213,466,917} - 1$, a 4,053,946-digit number. They use a program linked to the Great Internet Mersenne Prime Search (GIMPS), launched by Woltman in 1996, in which some 120,000 amateur and professional mathematicians are participating worldwide. However, Euclid had already proven 2,300 years ago that there is no end to the primes, so it is only a question of time before even this record will be broken.

SECTION FOUR
CHARTS & TABLES

A. TRIGONOMETRIC IDENTITIES

Signs of trigonometric functions

	Quadrant I	Quadrant II	Quadrant III	Quadrant IV
sin A	+	+	–	–
cos A	+	–	–	+
tan A	+	–	+	–
cot A	+	–	+	–
sec A	+	–	–	+
csc A	+	+	–	–

Values for special angles

A (degrees)	(radians)	sin A	cos A	tan A	cot A	sec A	csc A
0°	0	0	1	0	–	1	–
15°	$\pi/12$	$(\sqrt{6}-\sqrt{2})/4$	$(\sqrt{6}+\sqrt{2})/4$	$2-\sqrt{3}$	$2+\sqrt{3}$	$\sqrt{6}-\sqrt{2}$	$\sqrt{6}+\sqrt{2}$
30°	$\pi/6$	1/2	$\sqrt{3}/2$	$\sqrt{3}/3$	$\sqrt{3}$	$2\sqrt{3}/3$	2
45°	$\pi/4$	$\sqrt{2}/2$	$\sqrt{2}/2$	1	1	$\sqrt{2}$	$\sqrt{2}$
60°	$\pi/3$	$\sqrt{3}/2$	1/2	$\sqrt{3}$	$\sqrt{3}/3$	2	$2\sqrt{3}/3$
75°	$5\pi/12$	$(\sqrt{6}+\sqrt{2})/4$	$(\sqrt{6}-\sqrt{2})/4$	$2+\sqrt{3}$	$2-\sqrt{3}$	$\sqrt{6}+\sqrt{2}$	$\sqrt{6}-\sqrt{2}$
90°	$\pi/2$	1	0	–	0	–	1
120°	$2\pi/3$	$\sqrt{3}/2$	–1/2	$-\sqrt{3}$	$-\sqrt{3}/3$	–2	$2\sqrt{3}/3$
135°	$3\pi/4$	$\sqrt{2}/2$	$-\sqrt{2}/2$	–1	–1	$-\sqrt{2}$	$\sqrt{2}$
150°	$5\pi/6$	1/2	$-\sqrt{3}/2$	$-\sqrt{3}/3$	$-\sqrt{3}$	$-2\sqrt{3}/3$	2
180°	π	0	–1	0	–	–1	–
210°	$7\pi/6$	–1/2	$-\sqrt{3}/2$	$\sqrt{3}/3$	$\sqrt{3}$	$-2\sqrt{3}/3$	–2
225°	$5\pi/4$	$-\sqrt{2}/2$	$-\sqrt{2}/2$	1	1	$-\sqrt{2}$	$-\sqrt{2}$
240°	$4\pi/3$	$-\sqrt{3}/2$	–1/2	$\sqrt{3}$	$\sqrt{3}/3$	–2	$-2\sqrt{3}/3$
270°	$3\pi/2$	–1	0	–	0	–	–1
300°	$5\pi/3$	$-\sqrt{3}/2$	1/2	$-\sqrt{3}$	$-\sqrt{3}/3$	2	$-2\sqrt{3}/3$
315°	$7\pi/4$	$-\sqrt{2}/2$	$\sqrt{2}/2$	–1	–1	$\sqrt{2}$	$-\sqrt{2}$
330°	$11\pi/6$	–1/2	$\sqrt{3}/2$	$-\sqrt{3}/3$	$-\sqrt{3}$	$2\sqrt{3}/3$	–2
360°	2π	0	1	0	–	1	–

Relations among the trigonometric functions

In the following relations, the sign of each radical expression is determined by the quadrant in which the angle A lies.

	$\sin A = u$	$\cos A = u$	$\tan A = u$	$\cot A = u$	$\sec A = u$	$\csc A = u$
$\sin A$	u	$\pm\sqrt{1-u^2}$	$\pm u/\sqrt{1+u^2}$	$\pm 1/\sqrt{1+u^2}$	$\pm\sqrt{u^2-1}/u$	$1/u$
$\cos A$	$\pm\sqrt{1-u^2}$	u	$\pm 1/\sqrt{1+u^2}$	$\pm u/\sqrt{1+u^2}$	$1/u$	$\pm\sqrt{u^2-1}/u$
$\tan A$	$\pm u/\sqrt{1-u^2}$	$\pm\sqrt{1-u^2}/u$	u	$1/u$	$\pm\sqrt{u^2-1}$	$\pm 1/\sqrt{u^2-1}$
$\cot A$	$\pm\sqrt{1-u^2}/u$	$\pm u/\sqrt{1-u^2}$	$1/u$	u	$\pm 1/\sqrt{u^2-1}$	$\pm\sqrt{u^2-1}$
$\sec A$	$\pm 1/\sqrt{1-u^2}$	$1/u$	$\pm\sqrt{1+u^2}$	$\pm\sqrt{1+u^2}/u$	u	$\pm u/\sqrt{u^2-1}$
$\csc A$	$1/u$	$\pm 1/\sqrt{1-u^2}$	$\pm\sqrt{1+u^2}/u$	$\pm\sqrt{1+u^2}$	$\pm u/\sqrt{u^2-1}$	u

Complementary and supplementary relations (angles in degrees)

$\sin(90° - A) = \cos A$	$\cos(90° - A) = \sin A$	$\tan(90° - A) = \cot A$
$\cot(90° - A) = \tan A$	$\sec(90° - A) = \csc A$	$\csc(90° - A) = \sec A$
$\sin(90° + A) = \cos A$	$\cos(90° + A) = -\sin A$	$\tan(90° + A) = -\cot A$
$\cot(90° + A) = -\tan A$	$\sec(90° + A) = -\csc A$	$\csc(90° + A) = \sec A$
$\sin(180° - A) = \sin A$	$\cos(180° - A) = -\cos A$	$\tan(180° - A) = -\tan A$
$\cot(180° - A) = -\cot A$	$\sec(180° - A) = -\sec A$	$\csc(180° - A) = \csc A$
$\sin(180° + A) = -\sin A$	$\cos(180° + A) = -\cos A$	$\tan(180° + A) = \tan A$
$\cot(180° + A) = \cot A$	$\sec(180° + A) = -\sec A$	$\csc(180° + A) = -\csc A$
$\sin(270° - A) = -\cos A$	$\cos(270° - A) = -\sin A$	$\tan(270° - A) = \cot A$
$\cot(270° - A) = \tan A$	$\sec(270° - A) = -\csc A$	$\csc(270° - A) = -\sec A$
$\sin(270° + A) = -\cos A$	$\cos(270° + A) = \sin A$	$\tan(270° + A) = -\cot A$
$\cot(270° + A) = -\tan A$	$\sec(270° + A) = \csc A$	$\csc(270° + A) = -\sec A$
$\sin(360° - A) = -\sin A$	$\cos(360° - A) = \cos A$	$\tan(360° - A) = -\tan A$
$\cot(360° - A) = -\cot A$	$\sec(360° - A) = \sec A$	$\csc(360° - A) = -\csc A$
$\sin(360° + A) = \sin A$	$\cos(360° + A) = \cos A$	$\tan(360° + A) = \tan A$
$\cot(360° + A) = \cot A$	$\sec(360° + A) = \sec A$	$\csc(360° + A) = \csc A$

Basic identities

$\tan A = \sin A/\cos A \qquad \cot A = \cos A/\sin A \qquad \sec A = 1/\cos A \qquad \csc A = 1/\sin A$

Even-odd relations

$\sin(-A) = -\sin A$ $\qquad$ $\cos(-A) = \cos A$ $\qquad$ $\tan(-A) = -\tan A$

$\cot(-A) = -\cot A$ $\qquad$ $\sec(-A) = \sec A$ $\qquad$ $\csc(-A) = -\csc A$

Pythagorean relations

$\sin^2 A + \cos^2 A = 1$ $\qquad$ $1 + \tan^2 A = \sec^2 A$ $\qquad$ $1 + \cot^2 A = \csc^2 A$

Periodicity (angles in radians)

$\sin(A + 2\pi) = \sin A$ $\qquad$ $\cos(A + 2\pi) = \cos A$ $\qquad$ $\tan(A + \pi) = \tan A$

$\csc(A + 2\pi) = \csc A$ $\qquad$ $\sec(A + 2\pi) = \sec A$ $\qquad$ $\cot(A + \pi) = \cot A$

Addition formulas

$\sin(A + B) = \sin A \cos B + \cos A \sin B$

$\sin(A - B) = \sin A \cos B - \cos A \sin B$

$\cos(A + B) = \cos A \cos B - \sin A \sin B$

$\cos(A - B) = \cos A \cos B + \sin A \sin B$

$\tan(A + B) = (\tan A + \tan B)/(1 - \tan A \tan B)$

$\tan(A - B) = (\tan A - \tan B)/(1 + \tan A \tan B)$

$\cot(A + B) = (\cot A \cot B - 1)/(\cot A + \cot B)$

$\cot(A - B) = (\cot A \cot B + 1)/(\cot B - \cot A)$

Powers of trigonometric functions

$\sin^2 A = (1 - \cos 2A)/2$

$\cos^2 A = (1 + \cos 2A)/2$

$\sin^3 A = (3 \sin A - \sin 3A)/4$

$\cos^3 A = (3 \cos A + \cos 3A)/4$

Double-angle formulas

$\sin 2A = 2 \sin A \cos A$

$\cos 2A = \cos^2 A - \sin^2 A = 1 - 2 \sin^2 A = 2 \cos^2 A - 1$

$\tan 2A = 2 \tan A/(1 - \tan^2 A)$

$\cot 2A = (\cot^2 A - 1)/2 \cot A$

Half-angle formulas

In the following formulas, the sign of each radical is determined by the quadrant in which the angle A/2 lies.

$\sin A/2 = \pm\sqrt{(1 - \cos A)/2}$

$\cos A/2 = \pm\sqrt{(1 + \cos A)/2}$

$\tan A/2 = \pm\sqrt{(1 - \cos A)/(1 + \cos A)} = \sin A/(1 + \cos A) = (1 - \cos A)/\sin A = \csc A - \cot A$

$\cot A/2 = \pm\sqrt{(1 + \cos A)/(1 - \cos A)} = \sin A/(1 - \cos A) = (1 + \cos A)/\sin A = \csc A + \cot A$

Product-to-sum formulas

$\sin A \sin B = [\cos(A - B) - \cos(A + B)]/2$

$\cos A \cos B = [\cos(A - B) + \cos(A + B)]/2$

$\sin A \cos B = [\sin(A - B) + \sin(A + B)]/2$

Sum-to-product formulas

$\sin A + \sin B = 2 \sin[(A + B)/2] \cos[(A - B)/2]$

$\sin A - \sin B = 2 \cos[(A + B)/2] \sin[(A - B)/2]$

$\cos A + \cos B = 2 \cos[(A + B)/2] \cos[(A - B)/2]$

$\cos A - \cos B = -2 \sin[(A + B)/2] \sin[(A - B)/2]$

B. DIFFERENTIATION FORMULAS

General differentiation rules

In the following, f and g denote two differentiable functions of x, and c denotes a constant.

1. $d/dx\ cf(x) = c\ d/dx\ f(x)$
2. $d/dx\ [f(x) \pm g(x)] = d/dx\ f(x) \pm d/dx\ g(x)$
3. $d/dx\ f(x)g(x) = f(x)\ d/dx\ g(x) + g(x)\ d/dx\ f(x)$
4. $d/dx\ [f(x)/g(x)] = [g(x)\ d/dx\ f(x) - f(x)\ d/dx\ g(x)]/[g(x)]^2,\ g(x) \neq 0$
5. $d/dx\ f(g(x)) = f'(g(x))\ d/dx\ g(x) = f'(u)\ du/dx$, where $u = g(x)$

Derivatives

1. $d/dx\ c = 0$
2. $d/dx\ x^n = nx^{n-1}$
3. $d/dx\ e^x = e^x$
4. $d/dx\ \ln |x| = 1/x$
5. $d/dx\ \sin x = \cos x$
6. $d/dx\ \cos x = -\sin x$
7. $d/dx\ \tan x = \sec^2 x$
8. $d/dx\ \cot x = -\csc^2 x$

9. $d/dx\ \sec x = \sec x \tan x$
10. $d/dx\ \csc x = -\csc x \cot x$
11. $d/dx\ \arcsin x = 1/\sqrt{1-x^2}$
12. $d/dx\ \arccos x = -1/\sqrt{1-x^2}$
13. $d/dx\ \arctan x = 1/(1+x^2)$
14. $d/dx\ \text{arccot } x = -1/(1+x^2)$
15. $d/dx\ \text{arcsec } x = 1/|x|\sqrt{x^2-1}$
16. $d/dx\ \text{arccsc } x = -1/|x|\sqrt{x^2-1}$

17. $d/dx\ \sinh x = \cosh x$
18. $d/dx\ \cosh x = \sinh x$
19. $d/dx\ \tanh x = \text{sech}^2 x$
20. $d/dx\ \coth x = -\text{csch}^2 x$
21. $d/dx\ \text{sech } x = -\text{sech } x \tanh x$
22. $d/dx\ \text{csch } x = -\text{csch } x \coth x$

C. INTEGRATION FORMULAS

The following table gives indefinite integrals of the most commonly used elementary functions. For other indefinite integrals, consult a calculus textbook or a comprehensive mathematics reference book.

General integration rules

In the following, f and g denote two integrable functions of x, and C and k denote constants:

1. $\int kf(x)\ dx = k\int f(x)\ dx$
2. $\int [f(x) \pm g(x)]\ dx = \int f(x)\ dx \pm \int g(x)\ dx$
3. $\int f(x)\ g'(x)\ dx = f(x)\ g(x) - \int f'(x)\ g(x)\ dx$
4. $\int f(g(x))\ g'(x)\ dx = F(g(x)) + C$, where $F'(u) = f(u)$ and $u = g(x)$

Indefinite integrals (antiderivatives)

In the following, a and C denote a given constant and an arbitrary constant of integration, respectively:

1. $\int x^n \, dx = x^{n+1}/(n+1) + C, \; n \neq -1$

2. $\int x^{-1} \, dx = \int (1/x) \, dx = \ln |x| + C$

3. $\int 1/(x^2 + a^2) \, dx = (1/a) \arctan x/a + C$

4. $\int 1/(x^2 - a^2) \, dx = (1/2a) \ln |(x-a)/(x+a)| + C$
   $= -(1/a) \operatorname{arcoth} x/a + C \; (x^2 > a^2)$

5. $\int 1/(a^2 - x^2) \, dx = (1/2a) \ln |(x+a)/(x-a)| + C$
   $= (1/a) \operatorname{arctanh} x/a + C \; (x^2 < a^2)$

6. $\int 1/\sqrt{x^2 + a^2} \, dx = \ln |x + \sqrt{x^2 + a^2}| + C = \operatorname{arcsinh} x/a + C$

7. $\int 1/\sqrt{x^2 - a^2} \, dx = \ln |x + \sqrt{x^2 - a^2}| + C \; (x^2 > a^2)$

8. $\int 1/\sqrt{a^2 - x^2} \, dx = \arcsin x/a + C \; (x^2 < a^2)$

9. $\int 1/x\sqrt{x^2 + a^2} \, dx = -(1/a) \ln |[a + \sqrt{x^2 + a^2}]/x| + C$

10. $\int 1/x\sqrt{x^2 - a^2} \, dx = (1/a) \operatorname{arcsec} |x/a| + C \; (x^2 > a^2)$

11. $\int 1/x\sqrt{a^2 - x^2} \, dx = -(1/a) \ln |[a + \sqrt{a^2 - x^2}]/x| + C \; (x^2 < a^2)$

12. $\int e^{ax} \, dx = (1/a) e^{ax} + C$

13. $\int \ln x \, dx = x \ln |x| - x + C$

14. $\int (\ln x)/x \, dx = (1/2) \ln^2 |x| + C$

15. $\int \sin ax \, dx = -(1/a) \cos ax + C$

16. $\int \cos ax \, dx = (1/a) \sin ax + C$

17. $\int \tan ax \, dx = -(1/a) \ln |\cos ax| + C$

18. $\int \cot ax \, dx = (1/a) \ln |\sin ax| + C$

19. $\int \sec ax \, dx = (1/a) \ln |\sec ax + \tan ax| + C$

20. $\int \csc ax \, dx = (1/a) \ln |\csc ax - \cot ax| + C$

21. $\int \sin^2 ax \, dx = x/2 - (1/4a) \sin 2ax + C$

22. $\int \cos^2 ax \, dx = x/2 + (1/4a) \sin 2ax + C$

23. $\int \tan^2 ax \, dx = (1/a) \tan ax - x + C$

24. $\int \cot^2 ax \, dx = -(1/a) \cot ax - x + C$

25. $\int \sec^2 ax \, dx = (1/a) \tan ax + C$

26. $\int \csc^2 ax \, dx = -(1/a) \cot ax + C$

27. $\int e^{ax} \sin bx \, dx = e^{ax} (a \sin bx - b \cos bx)/(a^2 + b^2) + C$

28. $\int e^{ax} \cos bx \, dx = e^{ax} (a \cos bx + b \sin bx)/(a^2 + b^2) + C$

29. $\int \arcsin x/a \, dx = x \arcsin x/a + \sqrt{a^2 - x^2} + C \; (x^2 < a^2)$

30. $\int \arccos x/a \, dx = x \arccos x/a - \sqrt{a^2 - x^2} + C \; (x^2 < a^2)$

31. $\int \arctan x/a \, dx = x \arctan x/a - (a/2) \ln (x^2 + a^2) + C$

32. $\int \operatorname{arccot} x/a \, dx = x \operatorname{arccot} x/a + (a/2) \ln (x^2 + a^2) + C$

33. $\int \sinh ax \, dx = (1/a) \cosh ax + C$

34. $\int \cosh ax \, dx = (1/a) \sinh ax + C$

35. $\int \tanh ax \, dx = (1/a) \ln \cosh ax + C$

36. $\int \coth ax \, dx = (1/a) \ln |\sinh ax| + C$

37. $\int \operatorname{sech} ax \, dx = (2/a) \arctan e^{ax} + C$

38. $\int \operatorname{csch} ax \, dx = (1/a) \ln |\tanh ax/2| + C$

D. CONVERGENCE TESTS FOR SERIES

In the following, S is the sum of the series and L a finite limit.

Test	Series	Converges if	Diverges if	Remarks
nth term	$\sum_{n=1}^{\infty} a_n$		$\lim_{n \to \infty} a_n \neq 0$	
Geometric series	$\sum_{n=0}^{\infty} aq^n$	$\lvert q \rvert < 1$	$\lvert q \rvert \geq 1$	$S = a/(1-q)$
Telescopic series	$\sum_{n=1}^{\infty} (a_n - a_{n+1})$	$\lim_{n \to \infty} a_n = L$		$S = a_1 - L$
p-series	$\sum_{n=1}^{\infty} 1/n^p$	$p > 1$	$p \leq 1$	S depends on p.
Alternating series	$\sum_{n=1}^{\infty} (-1)^{n-1} a_n$	$0 < a_{n+1} \leq a_n$ and $\lim_{n \to \infty} a_n = 0$		
Integral (f continuous, positive, and decreasing)	$\sum_{n=1}^{\infty} a_n$ $a_n = f(n) \geq 0$	$\int_1^{\infty} f(x)\, dx$ converges	$\int_1^{\infty} f(x)\, dx$ diverges	
Root test	$\sum_{n=1}^{\infty} a_n$	$\lim_{n \to \infty} \sqrt[n]{\lvert a_n \rvert} < 1$	$\lim_{n \to \infty} \sqrt[n]{\lvert a_n \rvert} > 1$	test inconclusive if $\lim_{n \to \infty} \sqrt[n]{\lvert a_n \rvert} = 1$
Ratio test	$\sum_{n=1}^{\infty} a_n$	$\lim_{n \to \infty} \lvert a_{n+1}/a_n \rvert < 1$	$\lim_{n \to \infty} \lvert a_{n+1}/a_n \rvert > 1$	test inconclusive if $\lim_{n \to \infty} \lvert a_{n+1}/a_n \rvert = 1$
Direct comparison test ($a_n, b_n > 0$)	$\sum_{n=1}^{\infty} a_n$	$0 \leq a_n \leq b_n$ and $\sum_{n=1}^{\infty} b_n$ converges	$0 \leq b_n \leq a_n$ and $\sum_{n=1}^{\infty} b_n$ diverges	
Limit comparison test ($a_n, b_n > 0$)	$\sum_{n=1}^{\infty} a_n$	$\lim_{n \to \infty} a_n/b_n = L > 0$ and $\sum_{n=1}^{\infty} b_n$ converges	$\lim_{n \to \infty} a_n/b_n = L > 0$ and $\sum_{n=1}^{\infty} b_n$ diverges	

Based on: Roland E. Larson, Robert P. Hostetler, and Bruce H. Edwards, *Calculus with Analytic Geometry*, 5th ed. Lexington, Mass.: D.C. Heath, 1994, p. 594.

RECOMMENDED READING

Baron, Margaret E. *The Origins of the Infinitesimal Calculus.* New York: Dover Publications, 1987.

Beckmann, Petr. *A History of π.* Boulder, Colo.: The Golem Press, 1970.

Borowski, E. J., and J. M. Borwein. *Collins Dictionary of Mathematics,* 2d ed. Glasgow, U.K.: HarperCollins, 2002.

Boyer, Carl B. *The History of the Calculus and Its Conceptual Development.* New York: Dover Publications, 1959.

———. *A History of Mathematics,* 2d ed, revised by Uta C. Merzbach. New York: John Wiley & Sons, 1989.

Burton, David M. *History of Mathematics: An Introduction.* Dubuque, Iowa: Wm. C. Brown, 1995.

Connally, Eric, Deborah Hughes-Hallett, Andrew M. Gleason, et al. *Functions, Modeling, Change: A Preparation for Calculus.* New York: John Wiley & Sons, 2000.

Courant, Richard. *Differential and Integral Calculus,* 2 vols, 2d ed., trans. by E. J. McShane. London and Glasgow: Blackie & Son, 1956.

Eves, Howard. *An Introduction to the History of Mathematics,* 6th ed. Fort Worth, Tex.: Saunders College Publishing, 1992.

Hughes-Hallett, Deborah, Andrew M. Gleason, et al. *Calculus: Single Variable,* 2d ed. New York: John Wiley & Sons, 2000.

Katz, Victor J. *A History of Mathematics: An Introduction.* New York: HarperCollins, 1993.

Larson, Roland E., Robert P. Hostetler, and Bruce H. Edwards. *Calculus with Analytic Geometry,* 7th ed. New York: Houghton Mifflin, 2001.

Maor, Eli. *e: The Story of a Number.* Princeton, N. J.: Princeton University Press, 1994.

Nahin, Paul J. *An Imaginary Tale: The Story of $\sqrt{-1}$.* Princeton, N.J.: Princeton University Press, 1998.

Simmons, George F. *Calculus with Analytic Geometry,* 2d ed. New York: McGraw-Hill, 1995.

Stewart, James. *Single Variable Calculus: Early Transcendentals,* 4th ed. Pacific Grove, Calif.: Brooks/Cole, 1999.

Strauss, Monty J., Gerald L. Bradley, and Karl J Smith. *Calculus,* 3rd ed. Upper Saddle River, N.J.: Prentice Hall, 2002.

Thomas, George B., and Ross. L. Finney. *Calculus and Analytic Geometry,* 9th ed. Reading, Mass.: Addison-Wesley, 2000.

Toeplitz, Otto. *The Calculus: A Genetic Approach.* Edited by Gottfried Köthe, trans. by Luise Lange. Chicago & London: University of Chicago Press, 1981.

Wilson, Robin J. *Stamping Through Mathematics.* New York: Springer Verlag, 2001.

USEFUL WEBSITES

American Mathematical Society. http://www.ams.org/ Available on-line. Downloaded February 18, 2003.

Eric Weisstein's World of Mathematics. http://mathworld.wolfram.com/ Available on-line. Downloaded February 18, 2003.

High School Hub. http://highschoolhub.org/hub/math.cfm Available on-line. Downloaded February 18, 2003.

The Mathematical Association of America. http://www.maa.org/ Available on-line. Downloaded February 18, 2003.

Whatis.com. http://whatis.techtarget.com/definition/0,,sid9_gci803019,00.html Available on-line. Downloaded February 18, 2003.